W9-AUV-339

CONTEMPORARY MATHEMATICS

Titles in this Series

VOLUME 1 **Markov random fields and their applications**
Ross Kindermann and J. Laurie Snell

VOLUME 2 **Proceedings of the conference on integration, topology, and geometry in linear spaces**
William H. Graves, Editor

CONTEMPORARY MATHEMATICS

Volume 2

PROCEEDINGS OF THE CONFERENCE ON

Integration, Topology, and Geometry in Linear Spaces

AMERICAN MATHEMATICAL SOCIETY
Providence • Rhode Island

PROCEEDINGS OF THE CONFERENCE ON INTEGRATION, TOPOLOGY, AND GEOMETRY IN LINEAR SPACES

HELD AT THE UNIVERSITY OF NORTH CAROLINA
CHAPEL HILL, NORTH CAROLINA

MAY 17–19, 1979

EDITED BY
WILLIAM H. GRAVES

1980 Mathematics Subject Classifications. 28–02, 28A15, 28A20, 28A25, 28A33, 28B05, 28B10, 28C05, 46B22, 46G05, 46G10, 46E27, 46E40.

Library of Congress Cataloging in Publication Data

Conference on Integration, Topology, and Geometry in Linear Spaces, University of North Carolina, 1979.

Proceedings of the Conference on Integration, Topology, and Geometry in Linear Spaces, held at the University of North Carolina, Chapel Hill, May 17–19, 1979.

(Contemporary mathematics; v. 2)

Includes bibliographies.

1. Integrals, Generalized–Congresses. 2. Measure theory–Congresses. 3. Linear topological spaces–Congresses. 4. Pettis, Billy James. I. Graves, William Howard, 1940– II. Title. III. Series: Contemporary mathematics (Providence, R. I.); v. 2.

QA312.C575 1979 515.4'3 80-25417

ISBN 0-8218-5002-4

Copyright © 1980 by the American Mathematical Society
Printed in the United States of America
All rights reserved except those granted to the United States Government
This book may not be reproduced in any form without the permission of the publishers

CONTENTS

These Proceedings are dedicated to

B. J. PETTIS

1913–1979

INTRODUCTION

The papers in this collection are dedicated to the memory of Billy James Pettis who died of cancer at age 65 on April 14, 1979, shortly before his planned retirement as Professor of Mathematics, a position which he had held with distinction for 22 years at the University of North Carolina. His death barely preceded the Conference on Integration, Geometry, and Topology in Linear Spaces, which had been organized to coincide with his retirement. Professors Collins, Diestel, Huff, Kalton, Kluvanek, and Uhl spoke at the conference, and their contributions to this volume reflect their talks of May 17 and 18, 1979. Professor Dunford's paper represents the talk which he would have delivered had he not had to cancel his attendance at the last moment. The remaining papers herein were contributed by conference participants and, in the case of Professors Brooks and Dinculeanu, others who were forced to cancel their planned participation.

B. J. Pettis was raised in Spartanburg, South Carolina, where his father served Wofford College for many years as a teacher of mathematics and physics and his mother taught in the public schools.

He earned his B. A. from Wofford College in 1932, M. A. from the University of North Carolina in 1933, and Ph. D. from the University of Virginia in 1937. He then was Dupont Research Fellow at the University of Virginia in 1937–1938, Sterling Research Fellow at Yale in 1938–1939, and B. O. Peirce Instructor at Harvard from 1939 through early 1941 when he volunteered as a Private in the Army of the United States. After serving five years in the Army in Australia and Luzon and in combat in New Guinea and rising to the rank of Captain, he returned to mathematics. He served at Yale University (1945–1947), Tulane University (1947–1957, including one year as Chairman of the Department of Mathematics), Princeton University (1949–1950 as Visiting Lecturer), the National Science Foundation (1964–1965 as Science Faculty Fellow), Norfolk State College (1966–1977 as part time Visiting Professor), and, of course, the University of North Carolina (1957–1979).

Among his pre-war mathematical achievements were results which are now known to the mathematical world as the Orlicz-Pettis theorem, the Dunford-Pettis theorem, the Pettis theorem on measurability, and the theory of the Pettis integral. These alone would assure his place in mathematical history, but despite the professionally fracturing effect of serving in the Army, he returned from the war to write many widely read research articles and contribute to several books, including a now standard source: *Linear Topological Spaces* by Kelley, Namioka, and others. *Vector Measures* by Diestel and Uhl is dedicated to Bill Pettis. The authors of the latter note that "as we progressed in the study of the history of the basic theorems of the theory of vector measures, we were not surprised by learning that most of them, in one way or another, have their origins in the fertile mind of one man, B. J. Pettis." This dedication, a conference in 1975 on Pettis integration, and the conference represented by this volume attest to the value of his scientific life.

With his international reputation as a mathematical researcher and scholar came numerous invitations to serve his profession in broad and interesting ways. He was sought as a colloquium speaker and lecturer at universities, conferences, and institutes around the world. Leading mathematical journals such as the Bulletin of the American Mathematical Society and the Duke Journal sought and received his editorial services. He was a consultant in many capacities to many institutions, agencies, foundations, and societies including the National Science Foundation, the Woodrow Wilson Foundation, the Office of Naval Research, the Educational Testing Service, the American Mathematical Society, the Mathematical Association of America, the National Council of Teachers of Mathematics, the School Mathematics Study group, the Institute of International Education, numerous school systems, and several corporations, including the Education Development Corporation. It was in the service of the Education Development Corporation that over almost a decade he significantly contributed to the development of teacher training and mathematical curricula in eleven developing African countries as a participant and leader in several on-site institutes and writing projects. He held membership in many professional and honorary societies: the American Mathematical Society, the Mathematical Association of America, the Société Mathématique de France, Sigma Xi, the American Association for the Advancement of Science, the North Carolina Academy of Science, the National Council of Teachers of Mathematics, the American Association of University Professors, and Phi Beta Kappa.

No such formal listing of accomplishments can describe this remarkable man. He possessed a powerful yet graceful clarity of mind and expression. He greatly valued humbleness and good manners. That he could do so while holding himself, his colleagues, his friends, and his students to the highest standards is a tribute to his complete lack of pretension and his constant yet unobtrusive concern for the comfort and feelings of those around him. Some students will remember him as an inspirational teacher who taught them to discover their very best efforts within themselves. To many of his students he provided a first, and sometimes an only, encounter with depth and intellectual passion, and he will be remembered by them with a mysterious sense of respect. His ability and penchant for engaging others in gently probing, self-revealing conversation will be fondly remembered by those who knew him.

Bill Pettis loved his wife, Mary, his children, and all of his family. He immensely enjoyed the active appreciation of books, music, and wine which he shared with Mary, his family, and his friends. No theme more fascinated him than that of development. The development into graceful maturity of a young wine or a budding Ph. D. student afforded him a pleasure akin to that he derived and cherished from the development of his children. There was in him a mischievous, boyish quality which relished long evenings of wine and spirited conversation and hastily planned, devil-may-care visits to wine shops, book stores, and museums in distant cities. This impish aspect of his personality delighted his family and friends and complemented his graceful gentlemanly demeanor.

One of the finest assessments of the value of Bill Pettis to his colleagues was written to him by a friend when he learned of Bill's illness. "It wasn't too long after I'd come to know you that I felt 'There is a man to measure yourself by'. Not just things mathematical, for these the measure was obvious and woefully short on my end. But in all things. I felt in meeting you that I had finally come in contact with one of those special individuals, a kind I had only known before through time and distance on the printed page. You seemed a living link with those ideas and men who had fired me from as far back as I could remember. The clarity and depth of your expression remains my envy. I know you're no Newton or Jefferson or Gauss, but in my still boyish imagination I felt that in knowing you I had experienced some touch, however slight, with such as these."

He will long be remembered for his very positive influence on mathematics, on friends and colleagues, and on the shape of the Department of Mathematics at the University of North Carolina in Chapel Hill. He was a gentleman. He was a scholar. He was a man of worth.

William H. Graves
Robert L. Davis
Fred B. Wright

Contemporary Mathematics
Volume 2
1980

STRICT TOPOLOGIES IN MEASURE THEORY

Heron S. Collins

1. Introduction.

When I first was invited to take part in this conference on vector measures, knowing as I did that almost all the participants were Banach space theorists, my first reaction was that I didn't really fit in (spaces with strict topologies are never metrisable, much less Banachable). However, my deep affection for Billy Pettis, desire to pay some small tribute to him, and my knowledge of his predilection for soft analysis methods dictated that I participate. At the very least, I could point out that (i) strict topologies have been around for a long time, e.g., the topology bounded-weak * has played a large role in the study of completeness and weak compactness, (ii) many important spaces of scalar-valued measures which are found in probability theory are the duals of spaces of functions given strict topologies, and some interesting extensions of these have been made to vector-valued functions and measures, (iii) relatively recently, several mathematicians have made efforts (with some success) to systematize strict topology methods and language to a study of vector measures (and other areas) and, (iv) A. Grothendieck's work in the areas of interest of this conference used many techniques which, in retrospect, were strict topology methods.

Let me say at the outset that the phrase "strict topology methods" (as opposed to the by and large ad hoc methods applied to a particular strict topology, which so far have proved successful) may be a misnomer, for only in a few papers (see [3, 5, 7, 8, 22, 25, 26, and 29]) has there been any concerted effort to define and synthesize

Copyright © 1980, American Mathematical Society

the properties and methods here. The most ambitious efforts seem to me to be those in [3, 5, 25, 26, 7, 8].

2. Preliminaries.

We shall follow, so far as notation is concerned, the French school, as can be found in [17, 18, 24]. Thus if X is an ℓ.c.s. (locally convex space, and all spaces are Hausdorff), and Y another, $L(X,Y)$ denotes all continuous linear operators on X to Y, X' denotes the dual space of X, and $\widetilde{X}$ is the completion of X. On X' the topologies of main interest are $\sigma(X',X)$, $\beta(X',X)$, $\lambda(X',X)$, $\kappa(X',X)$, and $\tau(X',X)$, which are (resp.) the topologies of uniform convergence on finite sets, bounded sets, totally bounded sets, compact disks, and weakly compact disks. The first (in retrospect) strict topology that this writer encountered was bounded-weak * [10], and in [4] I defined and used its generalizations ew^* and cew^* to study completeness and weak compactness in ℓ.c.s. X, e.g., X is complete iff $(X', cew^*)' = X$ (here and afterward $(X,t)'$ denotes all t-continuous linear forms on X), X is fully complete (resp. has the Krein-Smulian property) iff each linear set (resp. convex set) of X' which is ew^* closed is already $\sigma(X',X)$ closed. The topology ew^* (resp. cew^*) is defined as the finest (resp. finest locally convex) topology on X' agreeing with $\sigma(X',X)$ on each equicontinuous set. Only when X is an (F) space with local base $\{U_n\}$ does the Banach-Dieudonné theorem [18] give the Krein-Smulian theorem or describe the relationship between ew^*, cew^*, $\lambda(X',X)$, and $\kappa(X',X)$: these topologies <u>all</u> coincide then, X has the Krein-Smulian property, and the sets $U_n^o = \{x' \in X' : |\langle x,x'\rangle| \leq 1$ for all $x \in U_n\}$ form a countable base for bounded sets in (X', cew^*) whose members are compact disks (thus, (X', cew^*) is a semi-Montel space).

When S is a (Hausdorff) completely regular space and X an ℓ.c.s., then $C(S,X)$ (resp. $C_o(S,X)$) denotes the space of all bounded continuous X-valued function on S (resp. those which also vanish at infinity). When X = scalars, the symbols became simply C(S), $C_o(S)$, and when S is locally compact the early work of Buck [1] (resp. Wells [31]) yielded a Riesz representation for $(C(S),\beta)'$ (resp. $(C(S,X),\beta)'$), where the strict topology β is defined by the seminorms $f \mapsto \sup\{P(\varphi(t)f(t)) : t \in S\}$, where P runs through a fundamental family of seminorms for X and φ runs through $C_o(S)$. The essential features of Buck's space $(C(S),\beta)$ are that β is coarser than the uniform topology yet yields the same bounded sets, gives the same dual as if S were compact (= all bounded Radon measures), is the finest ℓ.c.s. topology on C(S) agreeing with itself (or the compact-open topology) on bounded sets, and has a <u>countable</u> base for bounded sets (the last remark fails for $(C(S),X),\beta)$ unless X is at least an (F) space).

The above two examples (cew* and β) form the background for the definitions of strict topology we now give (see [3] for a first attempt at such a definition, and see also [7, 22, 25, 26]).

<u>2.1 Definition of strict topology</u>. An ℓ.c.s. (X, t) is a <u>strict space</u> with strict topology t if

2.1.1. t is the finest ℓ.c.s. topology agreeing with itself on t-bounded sets, and is a <u>(gDF) space</u> (for generalized (DF) space) if in addition

2.1.2. (X, t) has a countable base for bounded sets.

It should be remarked that our reason for using the term (gDF) in 2.1.2 (as opposed to the term D_b used in [22] for the same class of spaces) is due to our close association with W. Ruess who first proposed the name, and to our belief that Grothendieck, who invented the (DF) spaces [15], has contributed more than most to strict space techniques (without trying to do so!).

We now list some particular strict and (gDF) spaces, ones which seem to be relevant to this conference (see [3] for others).

3. Further examples of strict and (gDF) spaces.

3.1. The (DF) spaces of Grothendieck (and this class includes all normed spaces and strong duals of (F) spaces) are (gDF) spaces, by definition of (DF) space and [15,p.68], and there are (gDF) spaces which are not (DF) ; see 3.7.

3.2. In order to apply the tools of duality theory to probability and measure theory, Sentilles generalized Buck's example (given in 2 above) to completely regular (non-locally compact) spaces S and defined on $C(S)$ three strict topologies β_o, β, and β_1, whose duals yielded (resp.) the spaces $M_t(S)$ of tight Baire or Borel measures, $M_\tau(S)$ of τ-additive Baire measures, and $M_\sigma(S)$ of σ-additive Baire measures (see also [11]). In [32], Wheeler defined β_e on $C(S)$ and showed $(C(S),\beta_e)'$ is the space of separable measures of Dudley. All these topologies yield the same bounded sets as the uniform topology (as in Buck's case), and each makes $C(S)$ into a (gDF) space.

3.3. In [12], Fontenot defined on $C(S,X)$, S completely regular and X a normed space, three topologies β_o, β, β_1 (in analogy with 3.2): β_o is the finest l.c.s. topology agreeing with the compact-open topology on sup norm bounded sets, while the other two (just as in Sentilles' case) are inductive limits of topologies like β_o. In this paper and subsequent ones by Katsaras and Khurana [19,20,21] it is shown that these spaces are (gDF) spaces and that (roughly speaking, with some restrictions on β_1) the Riesz Representation type theorems quoted above in 3.2 go through: $(C(S,X),t)'$ is, for $t = \beta_o$, β, β_1, β_e, a suitable space of X'-valued measures. Not always does one need X to be normed (in which case one gets only a strict space). However, to obtain that the tensor product $C(S)\otimes X$ is dense in $(C(S,X),t)$, or that $(C(S,X),t)$ have the approximation property [16] or other desirable properties, one seems to need that the space $(C(S,X),t)$ be a (gDF) space.

3.4. It is known that the topology β_e is the finest ℓ.c.s. topology t on $C(S)$ which makes the embedding $f \to f^*$ from $C(S,X)$ into $X\varphi(C(S),t)$ continuous. Here, for X, Y ℓ.c.s., $X\varphi Y = L_e[(X',cew^*),Y]$, where the subscript e means "uniform convergence on equicontinuous sets of X' ", and for $f \in C(S,X)$, and $x' \in X'$, $f^*(x') = x' \circ f$. This fact led this author in [2] to an investigation of Waelbroeck's φ product (to see, for example, if interesting strict or other topologies arose on $C(S,X)$ under this embedding). More recently, in [5], this space $X\varphi Y$ played a key role in certain questions regarding weak compactness in, weak sequential completeness of, and reflexivity of certain spaces of operators. Here, $X\varphi Y$ turns out to be all hypocontinuous bilinear forms on $(X',cew^*) \times (Y',cew^*)$ and is thus the space of <u>continuous</u> bilinear forms when for example, X and Y are (F) spaces (see 4.9 below).

3.5. <u>The Saks spaces of J. B. Cooper</u>. In [7], there is developed a theory of special "mixed" spaces which arise as the finest ℓ.c.s. topology agreeing with a certain weaker topology on a given norm unit ball. These are the same as the <u>simple</u> D_b spaces of [22], and are those "one ball" (gDF) spaces in which a <u>single</u> bounded set absorbs all bounded sets. Of our previous examples, those of 3.2 are Saks spaces as are those of 3.3 if X is normed. Finally, the space (X',cew^*) is a Saks space when X is a Banach space. As is pointed out in [8] and as is well known, a σ-additive measure μ on a σ-field with values in a Banach space X takes its values in a weakly compact set, and the authors here fasten their attention on Saks spaces derived this way: if B is the closed disk generated by range μ , where μ maps Σ into an ℓ.c.s. (X,t), and X_B = linear span of B with norm given by the gauge of B , they consider the finest ℓ.c.s. topology on X_B agreeing with t on B , and succeed in obtaining certain results by Saks space methods.

3.6. Our next example is the universal strict topology of Graves [13]. Let R be a field of sets, $S(R)$ be all R-simple functions. On $S(R)$ there is a sup norm topology (just as in 3.2 above, and much use is made of this and other similarities to 3.2). Let τ be the finest ℓ.c.s. topology on $S(R)$ such that the (linearized) indicator function on R to $S(R)$ is strongly countably additive. If X is an ℓ.c.s. and μ a measure on R to X with $\tilde{\mu}$ the extension of $\hat{\mu}$ (where $\mu = \hat{\mu} \circ \chi$) over $\widetilde{S^{\tau}}(R)$, the completion of $(S(R),\tau)$, then (i) $\mu \to \tilde{\mu}$ is an isomorphism of s.c.a. (R,X) onto $L(\widetilde{S^{\tau}(R)},X)$, (ii) $S^{\tau}(R)$ and semi-reflexive $\widehat{S^{\tau}(R)}$ are (gdF) spaces which are Mackey when R is a σ-field, (iii) many of the results of measure theory follow here from what might be called strict topology techniques (soft analysis methods made possible by placing measure theory in the context of the universal strict topology). See also [14] for further results along these lines.

3.7. Finally, let E be an (F) space and let X be any one of the spaces $(E',\beta(E',E))$, $(E',\kappa(E',E))$, or $(E',\tau(E'E))$. Then the first is a (DF) space, hence (gDF), the second is the semi-Montel (gDF) space discussed in 2 above, and the last is a semi-reflexive (gDF) space. The last two are seldom (D F).

4. Some applications and properties of (gDF) spaces.

In [27], Sentilles derived a result about (weakly) compact operators on $(C(S),\beta)$, and in [25,29,8] these results are generalized by strict or Saks space methods, and Riesz representation theorems are obtained (see below in applications). Since these seem to be examples of "strict methods", we shall let the following theorems and their (sketched) proofs suffice for now (a later work [5] will be more revealing).

4.1. Theorem (Sentilles [27]). Let S,T be locally compact spaces, $u : (C(S),\beta) \to (C(T), \| \|_{\infty})$ be continuous and linear, and suppose u takes bounded sets to β (resp. β-weakly) relatively compact sets.

Then $u : (C(S),\beta) \to (C(T),\beta)$ is compact (resp. weakly compact).

Recall u on X to Y is (weakly) compact if it takes some zero neighborhood to a (weakly) relatively compact set.

The proof of 4.1 by Sentilles was quite complicated, using as it did kernel and measure theoretic techniques. The following theorems by Ruess [25], van Dulst [29] and Cooper and Schachermayer [8] are excellent examples of quite general results provable by abstracted properties and techniques. Ruess' result is more general than van Dulst's but we include both since van Dulst's proof is somewhat different. The Riesz - like theorem 4.4 generalizes those in [9].

4.2. Theorem (Ruess [25]). Let the $\ell.c.s.$ (X,τ) have properties 4.8 and 4.10 below and let (Y,ρ) be a quasi-complete $\ell.c.s.$ on which there is a metrisable topology $\rho_1 \geq \rho$. Then each $\tau-\rho_1$ continuous linear operator u on X to Y which takes τ-bounded sets to (weakly) relatively compact sets of (Y,ρ) is (weakly) compact on (X,τ) to (Y,ρ).

Proof. If $\{V_n\}$ is a ρ_1 zero neighborhood base consisting of closed disks for which $V_{n+1} \subset V_n$ for all n, then by hypothesis there is a sequence $\{a_n\}$ of positive scalars so that $U = \bigcap_{n=1}^{\infty} (a_n V_n)$ is also a zero neighborhood. By property 4.8 there is an equicontinuous $\sigma(X',X)$ closed disk $H \supset U^o$ such that the norm topology on X'_H with H as unit ball and $\beta(X',X)$ coincide on U^o. Since ρ is coarser than ρ_1, for any ρ zero neighborhood V there exists n so that $u'(V^o) \subset a_n U^o$. Following arguments like those of [17, P.179], one deduces that the polar of H in X is transformed by u into a (weakly) compact set of (Y,ρ).

4.3. Theorem (Van Dulst [29]). Suppose X satisfies 4.8 and 4.10 below and Y is an (F) space. Then each linear continuous operator u which takes bounded sets into (weakly) relatively compact sets is (weakly) compact.

Proof. Let $\{V_n\}$ be a base for zero neighborhoods consisting of closed (hence complete) disks. By 4.8 for each n there is a closed disk zero neighborhood U_n in X so that if $\varepsilon > 0$ there is bounded $M_{n,\varepsilon} \subset X$ such that $U_n \subset M_{n,\varepsilon} + \varepsilon u^{-1}(V_n)$. By 4.10 there is $\{a_n\}$ so $U = \bigcap_1^\infty a_n U_n$ is a zero neighborhood in X. Since $u(U) \subset a_n u(M_n, a_n^{-1}) + V_n$ all n, we know $u(U)$ is bounded. If we assume u takes bounded sets to relatively weakly compact sets, it will suffice by Alaoglu's theorem to show the $\sigma(Y'',Y')$ closure of $u(U)$ is contained in Y. But (– denotes $\sigma(Y'',Y')$ closure) $\overline{u(U)} \subset \bigcap_1^\infty [a_n u(M_n, a_n^{-1}) + \overline{V}_n]$ $\subset \bigcap_1^\infty (Y + \overline{V}_n) = Y$, completing the "weak" case, and the other follows even more quickly from $u(U) \subset a_n \cdot u(M_n, a_n^{-1}) + V_n$, all n, for then $u(U)$ is totally bounded.

A little notation is required for the following theorem, but not much, since we assume integration with respect to vector measures to be familiar (e.g., see [9] or [8]). If S is completely regular, $Bo(S)$ denotes the Borel sets of S (σ-field generated by closed sets), and a scalar-valued Radon measure on $Bo(S)$ is simply an element in $(C(S), \beta_o)'$, and an X-valued Radon measure is a bounded finitely additive inner regular measure on $Bo(S)$ to X (finite semivariation).

4.4. Theorem (Cooper-Schachermayer [8]). Let u be linear and continuous on $(C(S), \beta_o)$ to X (S completely regular and X an $\ell.c.s.$). Then there exists a Radon measure μ on $Bo(S)$ to $(X'', \sigma(X'', X'))$ such that $u(f) = \int f d\mu$ for all $f \in C(S)$. If u also maps bounded sets to relatively weakly compact sets, then μ takes its values in X and is Radon with respect to the original topology of X.

Proof. For the first assertion, let B = the $\sigma(X'',X')$- closure of the image under u of the closed unit ball of $C(S)$, so that B is a $\sigma(X'',X')$ compact disk (by Alaoglu's theorem). In the Banach space X''_B (= linear span of B with gauge of B as norm) there is a natural Saks space structure and strict topology: the finest on

X''_B agreeing with $\sigma(X'',X')$ on B, and B is compact in this topology. Let F be this Saks space, and regard u as a map from $(C(S),\beta_o)$ to F, so that by 4.2 above u is compact, and u'' on $M_t(S)'$ actually maps into F (as opposed to F''). If $A \in Bo(S)$ and χ_A is the indicator of A, then χ_A defines by integration an element of $M_t(S)'$ and so $\mu(A) = u''(\chi_A)$ is an F-valued measure and $u(f) = \int f d\mu$, all $f \in C(S)$.

For the second part, B as above lies in X already, and this completes the proof.

By using Grothendieck's characterization of weakly compact sets in $M_t(S)$, another result [17,p.94] of Grothendieck, Rosenthal's Lemma [9,I.4.1], and Saks space techniques, the above authors in [8] show that if the β_o continuous operator on $C(S)$ to any quasi-complete X fails to map bounded into relatively weakly compact sets then u fixes a subspace of $(C(S),\beta)$ which is isomorphic to c_o. As a result, if X fails to contain a copy of c_o (e.g., if X is wsc), then any β_o continuous linear operator on $C(S)$ to X must take bounded to weakly relatively compact sets (so, by 4.2 above, any such operator to an (F) space not containing c_o would have to be weakly compact). The authors also obtain by strict methods a rather nice proof of the spectral theorem for a Hermitian densely defined (not necessarily continuous) operator u on a Hilbert space. The spectrum $S = \sigma(u)$ of u is not compact and so on $C(S)$, the topology β_o works nicely.

4.5. <u>Some properties of (gDF) spaces</u>. Except for Grothendieck's completeness theorem and Wells' Riesz theorem (as remarked in 2 above), the restriction 2.1.2. above seems essential. We now list properties held by these (gDF) spaces, and throughout X denotes such (see [26] for proofs).

4.6. $L_b(X,Y)$ is complete for all complete ℓ.c.s. Y (b stands for the topology and uniform structure given by uniform convergence on bounded sets). In particular, $(X',\beta(X',X))$ is an (F) space.

4.7. A linear operator X to any ℓ.c.s. is continuous if it is continuous on bounded sets.

4.8. X is quasi-normable: given any zero neighborhood U in X there is another V such that $U^{o} \subset V^{o}$ and the norm topology of $X'_{V^{o}}$ restricted to U^{o} equals $\beta(X',X)$ restricted to U^{o}.

4.9. Hypocontinuous [16] bilinear forms on the product of two (gDF) spaces are continuous.

4.10. If $\{V_n\}$ is a sequence of closed disk zero neighborhoods in X there is a sequence $\{a_n\}$ of positive scalar so that $\bigcap_1^\infty a_n V_n$ is a zero neighborhood.

4.11. If X has basic sequence $\{A_n\}$ of bounded sets,

(i) $\tilde{X}$ is (gDF), with basic sequence $\{\tilde{A}_n\}$

(ii) if X is simple, with basic ball A, then $(\tilde{X},\tilde{A})$ is simple.

4.12. If $(X,\{A_n\})$, $(Y,\{B_n\})$ are (gDF) spaces, then

(i) $X \otimes_\pi Y$ and $X \tilde{\otimes}_\pi Y$ are (gDF), with basic sequences $\Gamma(A_n \otimes B_n)$ and $\widetilde{\Gamma(A_n \otimes B_n)}$ (projective and completed projective tensor products [16, 26],

(ii) (Problème des topologies [16]). Each bounded set of $X \otimes_\pi Y$ or $X \tilde{\otimes}_\pi Y$ is contained in some multiple of $\Gamma(A_n \otimes B_n)$ (resp.) $\widetilde{\Gamma(A_n \otimes B_n)}$.

(iii) If (X,A) and (Y,B) are simple semi-Montel (gDF) spaces then $X \tilde{\otimes}_\pi Y$ is also with $\widetilde{\Gamma(A \otimes B)}$ as one ball, and its bidual = $(X \tilde{\otimes}_\pi Y, \widetilde{\Gamma(A \otimes B)})$.

5. <u>Final comments</u>. Since the early work of Pettis [23], vector measures, tensor products, and integration techniques have played and are still playing a large role in the structure of Banach spaces. Among the properties encountered have been weak sequential completeness, reflexivity, Radon-Nikodym property, the approximation and

metric approximation properties, the Dunford-Pettis property, etc. In a later work [5], we hope to indicate how (DF) techniques may be used to unify and extend some known results and refinements of other results concerning some of these properties.

References

1. Buck, R. C., Bounded continuous functions on a locally compact space, Mch. J. Math. 5 (1958), 95-104.
2. Collins, H. S., Waelbroeck's φ product, unpublished manuscript.
3. Collins, H. S., Strict, weighted, and mixed topologies and applications, Adv. in Math. 19 (1976), 207-237.
4. Collins, H.S., Completeness and compactness in linear topological spaces, Trans. Amer. Math. Soc. 79 (1955), 256-280.
5. Collins, H. S. and Ruess, W., Weak compactness and weak convergence in spaces of operators, in preparation.
6. Conway, J. B., The strict topology and compactness in the space of measures, Trans. Amer. Math. Soc. 126 (1967), 474-486.
7. Cooper, J. B., Saks spaces and applications, Amsterdam, 1978.
8. Cooper, J. B. and Schachermayer, W., Saks spaces and vector-valued measures, Institutsbericht No. 98, Linz, 1978.
9. Diestel, J. and Uhl, J. J., Vector measures, mathematical surveys no. 15, Providence, 1977.
10. Eberlein, W. F., Weak compactness in Banach spaces I, P.N.A.S. 33(1947),51-53.
11. Fremlin, D. H., Garling, D. J. H., and Haydon, R. G., Bounded measures on topological spaces, Proc. London Math. Soc. 25 (1972), 115-136.
12. Fontenot, R. A., Strict topologies for vector-valued measures, Can. J. Math. 26 (1974), 841-853.
13. Graves, W. H., On the theory of vector measures, Mem. Amer. Math. Soc. no. 195, Providence, 1977.

14. Graves, W. H. and Sentilles, Dennis, The dual of the space of measures and completion of the universal measure, to appear J. Math. Anal. and Appl.

15. Grothendieck, A., Sur les espaces (F) et (DF), Summa Brasil. Math. 3 (1954), 57-122

16. Grothendieck, A., Produits tensoriels topologiques et espaces nucléaires, Mem. Am. Math. Soc. 16, Providence 1955.

17. Grothendieck, A., Topological vector spaces, Gordon and Breach, New York, 1973.

18. Horváth, J., Topological vector spaces and distributions, vol. 1, Addison-Wesley, Reading, 1966.

19. Katsaras, A. K., Spaces of vector measures, Trans. Amer. Math. Soc. 206 (1975), 313-328.

20. Katsaras, A. K., On the strict topology in the locally convex setting, Math. Ann. 216 (1975), 105-112.

21. Khurana, S. S., Topologies on spaces of vector-valued continuous functions, to appear Trans. Amer. Soc.

22. Noureddine, K., Note sur les espaces D_b , Math. Ann. 219, (1976), 97-103.

23. Pettis, B. J., On integration in vector spaces, Trans. Amer. Math. Soc. 44 (1938), 277-304.

24. Robertson, A. and Robertson, W., Topological vector spaces, Cambridge, 1964.

25. Ruess, W., On the locally convex structure of strict topologies, Math. Z. 153 (1977), 179-192.

26. Ruess, W., Halbnorm-dualität und induktive limestopologien in der theorie localkonvexer Räume, Habilitationschrift, Bonn, 1976.

27. Sentilles, Dennis, Compact and weakly compact operators on $(C(S),\beta)$, Ill. J. Math. 13 (1969), 769-776.

28. Sentilles, Dennis, Bounded continuous functions on a completely regular space, Trans. Amer. Math. Soc. 168 (1972), 311-336.

29. Van Dulst, D., (Weakly) compact mappings into (F) spaces, Math. Ann. 224 (1976), 112-116.

30. Waelbroeck, L., Duality and the injective tensor product, Math. Ann. 163 (1966), 122-126.

31. Wells, J., Bounded continuous vector-valued functions on a locally compact space, Mich. J. Math. 12 (1965), 119-126.

32. Wheeler, R. F., The strict topology, separable measures, and paracompactness, Pac. J. Math. 47 (1973), 287-302.

Department of Mathematics
Louisiana State University
Baton Rouge, La. 70803, U.S.A.

Contemporary Mathematics
Volume 2
1980

A SURVEY OF RESULTS RELATED TO THE DUNFORD-PETTIS PROPERTY

by Joe Diestel*

Dedicated to the memory of
Professor B. J. Pettis, a true
gentleman and scholar.

As you might guess, our story starts with the now classical Transactions paper [DP] of Nelson Dunford and Bill Pettis, "Linear Operations on Summable Functions." The present survey hopes to convey the spirit and some latter-day consequences of (generalizations of) two of the many gems to be found in the Dunford-Pettis classic. These two results will be referred to simply as

The Dunford-Pettis Theorem. *For any μ and any Banach space X, if $T: L_1(\mu) \to X$ is a weakly compact linear operator, then T is completely continuous.*

For any μ and any separable dual space Y, any linear operator $T: L_1(\mu) \to Y$ is completely continuous.

Despite several striking applications of the above theorem already found in the Dunford-Pettis paper, its results were to lay in rest for more than a decade. Finally, in the early 1950's Grothendieck launched his monumental study of Banach spaces and, in his ever-present homological bent of mind, isolated large classes of Banach spaces by means of what special classes of operators did to them. In considering the first assertion of the Dunford-Pettis Theorem, he canonized those Banach spaces which share with $L_1(\mu)$'s the property that weakly compact operators are completely

*The research conducted during the writing of this paper was supported by the National Science Foundation.

Copyright © 1980, American Mathematical Society

continuous "spaces with the Dunford-Pettis property." His remarkable "Sur les applications lineaires faiblement compactes d'espaces du type C(K)" is the birthplace of the property to which this survey, in the main, addresses itself.

The second assertion is, as we all know, a consequence of a more penetrating fact regarding separable dual spaces: they have the Radon-Nikodym property. However, in keeping with the spirit of this survey it is the consequence (namely the second assertion of the Dunford-Pettis Theorem) rather than the more penetrating fact which will occupy our attention here. It seems only fair, however, to mention that with a bit of work and the magic of the W. J. Davis, T. Figiel, W. B. Johnson and A. Pelczynski [DFJP] factorization theorem, the second assertion of the Dunford-Pettis theorem can be used to prove the first.

Incidentally, for those who like happy endings to stories we recommend Jean Bourgain's paper [JB2] where he shows that the Radon-Nikodym property for Y is equivalent to the representability of each completely continuous operator from L^1 to Y thereby closing the story of the Dunford-Pettis theorem in a most enjoyable manner.

This survey then hopes to discuss the Dunford-Pettis theorem cited above, most particularly the first assertion. We'd initially hoped to give an update on the Grothendieck paper with an extensive discussion of reciprocal Dunford-Pettis properties; however, trips down interesting side paths, lack of sufficient knowledge and even insufficient stamina forced abandonment on such a task. So we have, principally, concentrated on the Dunford-Pettis property. At no time have we shied away from diversions and we hope a few of these will find some interest. We have shied away (with but a few exceptions) from proving anything about special spaces. To prove completely that C(K) spaces, L_1-spaces, the disk algebra on various uniform algebras have the Dunford-Pettis property would have been a dubious service at best since well exposed proofs exist already; certainly complete proofs would have more than doubled the length of this report. Instead we have concentrated on presenting only the softest of the soft proofs preferring to keep our cannons in the references.

With evident self-interest we mention that many of the topics discussed (even in passing) herein are given a somewhat more extensive treatment in the "Sequences and Series in Banach Spaces" notes of the author. In particular Rosenthal's ℓ_1 theorem and Elton's c_0-characterization are presented therein with full Ramsey calisthenics beforehand.

Throughout this survey I've placed open problems. These are no doubt known to Dunford-Pettis fanatics everywhere. When possible I've visited upon the instigator of a given problem blame. I am indebted to these people for much of my interest in the subject. I'd also like to take this opportunity to thank Professor Graves for his incredible patience with my extreme tardiness with this report; it is much appreciated and I am indebted. Finally, special thanks go to Mrs. Julie Froble for her expert typing of this manuscript and to the many fellow mathematicians who've answered my questions, contributed examples (used and unused) and filled in gaps in knowledge.

The Dunford-Pettis Property

A Banach space X is said to have the Dunford-Pettis property if for each Banach space Y every weakly compact linear operator T: $X \to Y$ is completely continuous, i.e., T takes weakly compact sets in X onto norm compact sets in Y.

Our first result summarizes the best known (and a few not-so-well-known) equivalent formulations of the Dunford-Pettis property.

Theorem 1. *Each of the following conditions is equivalent to the condition that a Banach space X have the Dunford-Pettis property.*

(a) *For all Banach spaces Y, every weakly compact operator from X to Y sends weakly convergent sequences into norm convergent sequences.*

(b) *For all Banach spaces Y, every weakly compact operator from X to Y sends weakly Cauchy sequences into norm convergent sequences.*

(c) *If (x_n) and (x_n^*) are weakly null sequences in X and X^* respectively then $\lim_n x_n^* x_n = 0$.*

(d) *If (x_n) is a weakly Cauchy sequence in X and (x_n^*) is a weakly null sequence in X^* then $\lim_n x_n^* x_n = 0$.*

(e) *For all Banach spaces Y, every operator from X to Y with almost weakly compact adjoint is completely continuous.*

(f) *If (x_n) is a weakly null sequence in X and (x_n^*) is weakly Cauchy in X^*, then $\lim_n x_n^* x_n = 0$.*

(g) *Every operator T: $X \to c_0$ with almost weakly compact adjoint is completely continuous.*

(h) *For all Banach spaces Y, every almost weakly compact operator T: $Y \to X$ has a completely continuous adjoint.*

Many of the proofs involved in demonstrating Theorem 1 are virtually identical so we will touch on only a couple of the implications; perhaps the only other word-to-the-wise that has to be passed on is the definition of an almost weakly compact operator: the operator T: $X \to Y$ is almost weakly compact **if** for each bounded sequence (x_n) in X, (Tx_n) has a weakly Cauchy subsequence.

As examples of how to prove most all of the implications above we'll show that a space X with the Dunford-Pettis property has (c) and that (c) implies (d).

Suppose X has the Dunford-Pettis property and let (x_n) and (x_n^*) be as in (c). Define G: $\ell_1 \to X$ and A: $X \to c_0$ by $G(\lambda_n) = \Sigma_n \lambda_n x_n$, $Ax = (x_n^* x)$. Since G takes the closed unit ball of ℓ_1 into the absolutely closed convex hull of $\{x_n\}$, G is weakly compact; since A^*: $\ell_1 \to X^*$ is easily seen to be of the form $A^*(\lambda_n) = \Sigma_n \lambda_n x_n^*$, A^*, and hence A, is also weakly compact. Since X has the Dunford-Pettis property, AG: $\ell_1 \to c_0$ is compact. This and a quick computation now shows that $0 = \lim_n x_n^* x_n$.

Suppose you know of (c). If (y_n) is a weakly Cauchy sequence in X and (y_n^*) is weakly null in X^* yet $(y_n^* y_n)$ doesn't converge to zero then by passing to a subsequence if necessary we may assume that $|y_n^* y_n| \geq \delta > 0$ for some $\delta > 0$ and all n. Now (y_n^*) being weakly null implies there's a subsequence $(y_{k_n}^*)$ of (y_n^*) for which $|y_{k_n}^*(y_n)| \leq \delta/2$ for all n. Writing $y_{k_n}^* y_{k_n}$ as $y_{k_n}^*(y_{k_n} - y_n) + y_{k_n}^*(y_n)$ we see that $(y_{k_n} - y_n)$ is weakly null; it follows from (c) that for n large enough $|y_{k_n}^*(y_{k_n} - y_n)| \leq \delta/4$. But now for such n's:

$$\delta \leq |y_{k_n}^* y_{k_n}| \leq |y_{k_n}^*(y_{k_n} - y_n)| + |y_{k_n}^*(y_n)| \leq \frac{3\delta}{4}.$$

Oops.

To be sure, the above formulations are of interest principally because of the conclusions drawn from the Dunford-Pettis property; in all cases, the proofs of equivalence must be considered largely formal with condition (c) being (perhaps) the most easily tested. Conditions (a) through (c) were shown equivalent to the Dunford-Pettis property by J. Brace and A. Grothendieck (see p. 177 of [ ]); the equivalence of (d) and (f) to the rest is implicitly contained in the Brace-Grothendieck results; the rest of the equivalences might first have been noticed by H. Fakhoury [ ].

A quick consequence of Theorem 1(c) is

Corollary 2. *If X^* has the Dunford-Pettis property then so does X.*

Of course, much of the interest in the Dunford-Pettis property derives from the importance of the special spaces that enjoy the property. Which are these? To start with we have the Dunford-Pettis theorem itself telling us that *all L_1 spaces have the Dunford-Pettis property*. As noted in the Introduction, Dunford and Pettis proved most of this result; in fact, they showed that if T is a weakly compact linear operator on $L_1(\mu)$ for μ a finite measure and if T has separable range then T is completely continuous. R. S. Phillips [Ph] noticed that if T is a weakly compact operator on $L_1(\mu)$ for μ σ-finite, then the range of T is necessarily separable. Finally, it is easy to verify that if μ is any measure and (f_n) is a sequence in $L_1(\mu)$, then there is a sub-σ-field Σ_0 of μ's domain such that each f_n is Σ_0 measurable, $\mu|_{\Sigma_0}$ is σ-finite and $L_1(\Sigma_0, \mu|_{\Sigma_0})$ is a closed linear subspace of $L_1(\mu)$ on which T acts weakly compactly; in tandem with Theorem 1(a) this allows a quick proof of the full-fledged Dunford-Pettis Theorem.

Now it's well known that if K is a compact Hausdorff space then the dual $C(K)^*$ of C(K) is the space M(K) of regular Borel measures on K with variation norm. But M(K) is itself an L_1 space; it follows then from Corollary 2 that *all C(K) spaces have the Dunford-Pettis property*.

Curiously, for some time the only known examples of spaces with the Dunford-Pettis property were C(K) spaces, L_1-spaces and spaces whose duals were either C(K) or L_1 spaces. In fact, for a long time it was not known whether X's having the Dunford-Pettis property was equivalent to X^*'s having it. Such was put to rest by C. Stegall who noticed the following: if X is the ℓ_1-sum of n-dimensional Euclidean spaces, then X has the Schur property (weakly convergent sequences in X are norm convergent). By Theorem 1(i) it is plain that *spaces with the Schur property have the Dunford-Pettis property*. On the other hand, Charles Stegall [Sℓ_1] has proved the following fact useful for purposes of rather building spaces whose duals contain prescribed complemented subspaces:

The Principle of Local Selections. *Let T: $X \to Y$ be a bounded linear operator and suppose that Y has the bounded approximation property. Suppose that there is a $\lambda \geq 1$ such that for any finite dimensional Banach space Z any operator S: $Z \to Y$ and any $\epsilon > 0$ there is an operator $\hat{S}$: $Z \to X$ with $\|\hat{S}\| \leq \lambda\|S\|$ such that $\|T\hat{S} - S\| \leq \epsilon$.*

Then T is a quotient map whose adjoint's range is a complemented subspace of X^*.

A word or two about the idea of this principle may be in order, particularly since the statement is loaded with so many quantifiers. In fact the idea is really quite simple. What makes T^*Y^* complemented in X^*? A moment's reflection tells you that the existence of a bounded linear projection onto T^*Y^* is equivalent to the existence of a bounded linear operstor $S: X^* \to Y^*$ for which $ST^* = id_{Y^*}$. How to build such an S? Well, in the case at hand, we want S to achieve the identity $ST^* = id_{Y^*}$. Since Y has the bounded approximation property we have the existence of a uniformly bounded set of finite rank operators $I_\alpha: Y \to Y$ converging strongly to id_Y; hand-in-hand with this the I_α^*'s exhibit a type of convergence to $id_{Y^*} = id_Y^*$: if $y \in Y$, $y^* \in Y^*$, then $(I_\alpha^* y^*)(y) \to y^*(y)$. The point of assuming that T admits local selections is that by looking at $I_\alpha T$: $I_\alpha T(Y) \to Y$ we can "lift" to operators $I_\alpha T$: $I_\alpha T(Y) \to X$ in a λ-controlled fashion and nearly achieve the identity $T\, I_\alpha T = I_\alpha$ (in duality: $I_\alpha^* = (I_\alpha T)^* T^*$). Hopefully by taking appropriate limits the "nearly"'s will pass away and will end up with $id_{Y^*} = \lim_\alpha (I_\alpha T^*) T^*$; whatever $\lim_\alpha I_\alpha T^*$ might be, it's clearly our best candidate for the operator S.

Now for the details:

<u>Proof</u>. Let $I_\alpha: Y \to Y$ $(\alpha \in D)$ be a net of finite rank operators on Y, converging to id_Y strongly for which $\|I_\alpha\| \leq M$ for all $\alpha \in D$ and let $J_\alpha: I_\alpha(Y) \hookrightarrow Y$ denote the natural inclusion maps of the finite dimensional subspaces $I_\alpha(Y)$ of Y into Y. For each $\alpha \in D$ there is an operator $\hat{J}_\alpha: I_\alpha Y \to X$ with $\|\hat{J}_\alpha\| \leq \lambda$ and $\|T\cdot\hat{J}_\alpha - J_\alpha\| < 1/[1 + \dim I_\alpha(Y)]^2$. Now $(\hat{J}_\alpha I_\alpha: Y \to X)_{\alpha\in D}$ is a uniformly bounded net of operators from Y to X. Looking at the net of adjoints: $(\hat{J}_\alpha I_\alpha)^*_{\alpha\in D} = (I_\alpha^* \hat{J}_\alpha^*)_{\alpha\in D}$, we get a net of operators from X^* to Y^* whose norms are uniformly bounded by $M\lambda$. By standard weak* compactness arguments there is a cluster point S in the topology of pointwise convergence on $E^* \times F$ (or, if you are so moved, $(\hat{J}_\alpha I_\alpha)^*_{\alpha\in D}$ has a cluster point S in $\mathcal{L}(X^*; Y^*)$ relative to the weak* topology generated by the projective tensor product $X^* \hat{\otimes} Y$ of which $\mathcal{L}(X^*; Y^*)$ is the dual!). Our claim is that ST^* is the identity operator

on Y^*; once established T^*S will be a bounded linear projection of X^* onto Y^*.

Let $y \in Y$ and $y^* \in Y^*$. Take note of the following salient features of the above discussion.

1. $I_\alpha y \to y$, as α goes in direction D; so $y^*T_\alpha y \to y^*y$.

2. $\|(T\hat{J}_\alpha)^* - J_\alpha^*\| = \|T\hat{J}_\alpha - J_\alpha\|$ gets as close to zero as you please merely by sending α out along direction D very far.

3. $y \in Y$ and $T^*y^* \in X^*$, so if α is very far out in direction D we have $S(T^*y^*)(y)$ close to $(\hat{J}_\alpha I_\alpha)^*(T^*y^*)(y)$.

Putting these features together and letting α go very, very far in the direction D we see (using $\sim$ to mean "is close to") that

$$\begin{aligned} y^*y \sim y^*I_\alpha y \sim (I_\alpha^* y^*)(y) \\ \sim (\hat{J}_\alpha I_\alpha)^*(y^*)(y) \sim (T\hat{J}_\alpha I_\alpha)^*(y^*)(y) \\ \sim (\hat{J}_\alpha I_\alpha)^*(T^*y^*)(y) \\ \sim S(T^*y^*)(y) = (ST^*y^*)(y). \end{aligned}$$

The upshot of all this is that for any $y \in Y$, $y^* \in Y^*$, $(ST^*)(y^*)(y) = id_{Y^*}(y^*)(y)$ and this is as it should be.

Back to the issue at hand. Letting X be the Schur space $(\Sigma \oplus \ell_2^n)_{\ell_1}$, Y be the space ℓ_2 and T be the operator from X to Y defined by taking a typical vector from $(\Sigma \oplus \ell_2^n)_{\ell_1}$, say

$$((r_{1,1}), (r_{2,1}, r_{2,2}), (r_{3,1}, r_{3,2}, r_{3,3}), \cdots, (r_{n,1}, r_{n,2}, \cdots, r_{n,n}), \cdots)$$

to the member $(r_{1,1} + r_{2,1} + \cdots, r_{2,2} + r_{3,2} + \cdots, \cdots, r_{n,n} + r_{n,n+1} + \cdots, \cdots)$ of ℓ_2, we get that T admits local sections and so X^* admits a complemented copy of ℓ_2. Were X^* to have the Dunford-Pettis property as well, then the projection P of X^* onto ℓ_2 would be weakly compact (since ℓ_2 is reflexive) have compact square (= P itself!) making ℓ_2 finite dimensional. Consequently, $(\Sigma \oplus \ell_2^n)_{\ell_1}$ has the Dunford-Pettis property while its dual $(\Sigma \oplus \ell_2^n)_{\ell_\infty}$ does not.

We'll return to Stegall's example a bit later. It is worth mentioning that the approximation assumption present in the Principle of Local Selections can be eliminated if one uses a Lindenstrauss compactness argument. This improvement is due to W. B. Johnson [WBJ] and its statement reads as follows:

The Principle of Local Selections (Revisited). *Suppose T is a bounded linear operator from X to Y and let* $(F_\alpha)_{\alpha\in D}$ *be a net of subspaces of Y, directed by inclusion, with* $Y_0 = \bigcup F_\alpha$ *dense in Y. Assume that for each* α *there is an operator* $L_\alpha: F_\alpha \to X$ *such that* $TL_\alpha = id_{F_\alpha}$ *and* $\lim\sup_\alpha \|F_\alpha\| \leq \lambda < \infty$. *Then T is a quotient operator from X onto Y whose adjoint's range is complemented in* X^* *by a projection of norm* $\leq \lambda \|T\|$.

Using this version of Stegall's Principle, it is not difficult to show that *if Y is any separable Banach space then there exists a Banach space X with the Schur property such that* X^* *contains a complemented copy of* Y^*. We refer the reader to Johnson's paper for the ideas necessary to prove this.

At present no reasonable condition seems to be known which added to the assumption that X has the Dunford-Pettis property implies X^* has it too. The best that can be said (and it is clearly overkill) is the following:

Theorem 3: *If X has the Dunford-Pettis property and contains no copy of* ℓ_1 *then* X^* *has the Schur property and, therefore, the Dunford-Pettis property as well.*

The proof of this theorem is implicitly in H. P. Rosenthal [Rinj] and explicitly in H. Fakhoury [F] and Pethe and Thakare [PT]; it follows so easily from (d) in Theorem 1 and the ever present Rosenthal ℓ_1-theorem that we include the proof below. It is perhaps noteworthy that *X^*'s being a Schur space implies that X has the Dunford-Pettis property and that X contains no copy of* ℓ_1. In particular, the additional assumption that X contain no copy of ℓ_1 leaves a lot of fat to be trimmed before "reasonable" additional assumptions are found that allow one to conclude from X's having the Dunford-Pettis property that X^* has it too. A starting point might be

Question 1. If X has the Dunford-Pettis property and contains no complemented copy of ℓ_1 then need X^* have the Dunford-Pettis property?

Even should Question 1 have an affirmative answer there'd still be a lot of loose ends in the duality of Dunford-Pettis to tie up. It might even be that the key condition sought after be more "local" in nature, i.e., depend on the finite dimensional subspace structure of the space X. More on this and further examples of spaces with the Dunford-Pettis property will be discussed later.

Proof of Theorem 3. Suppose $X \not\supset \ell_1$ and X^* fails the Schur property; let (x_n^*) be a sequence of norm-one elements of X^* tending weakly to zero. Pick x_n in X so that $\|x_n\| = 1$ and $x_n^* x_n > 1/2$. By Rosenthal's ℓ_1-theorem there's a subsequence of (x_n) that's weakly Cauchy. Relabeling, if necessary, we now have (x_n) weakly Cauchy in X and a weakly null sequence (x_n^*) in X^* with $x_n^* x_n > 1/2$. This can't happen in a space X with the Dunford-Pettis property.

Despite the early paucity of examples, the importance of the Dunford-Pettis property was quick to gain recognition; at least as quick, that is, as the work of Grothendieck in functional analysis came to be understood. A few illustrative examples of the role played (or potentially played) by the Dunford-Pettis property are in order.

Recall that in the early 1950's the approximation problem was still open (the same could be said until the early 1970's, too). In his efforts to resolve this question, Grothendieck showed that an affirmative solution to the approximation problem for all Banach spaces follows from the possession of the approximation property by all subspaces of c_0; this is an easy consequence, by the way, of the quickly derived fact that every compact linear operator between two Banach spaces factors compactly through a subspace of c_0. This reduction to subspaces of c_0 lead Grothendieck to study the finer structure of such spaces. In short order he was able to show that every closed linear subspace of c_0 has the Dunford-Pettis property. Let's see why this is. Let X be a closed linear subspace of c_0, let (x_n) be a weakly null sequence in X and suppose (x_n^*) is a weakly null sequence in X^*. Suppose that for infinitely many n's, $x_n^*(x_n) \geq \delta > 0$; we may as well assume $\|x_n\| \geq \delta > 0$ for such n's. Start with x_1. Choose a coordinate k_1 such that x_1's norm is achieved in the indices before k_1. Use the fact that (x_n) is a weakly null sequence to see that for some $N_1 \geq 1$ if $n \geq N_1$ then $|x_n(1)|, \cdots, |x_n(k_1)| \leq \frac{\delta}{8}$. Pick $n_2 > N_1$ and choose $k_2 > k_1$ so that x_{n_2}'s norm is achieved in its coordinates prior to k_2. Use (x_n)'s weakly null nature once more to find $N_2 > n_2$ for which if $n \geq N_2$ then $|x_n(1)|, \cdots, |x_n(k_2)| \leq \frac{\delta}{8}$. Pick $n_3 > N_2$ and choose $k_3 > k_2$ so that k_{n_3}'s norm is achieved in its coordinates prior to k_3. Continue; the result is a small perturbation of a sequence (σ_n) of vectors in c_0 of the form $\sigma_j = \sum_{i=k_{j-1}+1}^{k_j} x_j(i)e_i$, where e_n is the nth unit vector in c_0. Now it is an easy matter to show that the sequence (σ_n) is a Schauder basis for its closed linear span that is, in fact, equivalent to the unit vector basis of c_0; small perturbations of such are also equivalent to the unit vector basis of c_0. So for some appropriate strictly increasing sequence (k_n) of positive integers, (x_{k_n}) is equivalent to the unit vector basis of c_0 and, in particular, the closed linear span $[x_{k_n}]$ of the x_{k_n}'s is isomorphic to c_0. Let's return to the x_n^*'s: (x_n^*) is still weakly null in X^* hence the sequence $(x_{k_n}^*|_{[x_{k_n}]})$ of restrictions of the $x_{k_n}^*$'s to $[x_{k_n}]$ is also weakly null.

But $[x_{k_n}]$ is isomorphic to c_0, so $(x_{k_n}^*|_{[x_{k_n}]})$ can be viewed as a sequence in ℓ_1; by Schur's theorem,

$$\|x^*_{k_n}|[x_{k_n}]\|_{\ell_1} \to 0.$$

In particular, we get $x^*_{k_n} x_{k_n}$ tending to zero. Oops.

Remark: The attentive reader will notice that in the above we've essentially argued to the Bessaga- Pelczynski Selection Principle [BP1]: *if (x_n) is a sequence in a Banach space that tends to zero weakly but not in norm, then (x_n) has a subsequence (y_n) which is a Schauder basis for its closed linear span*. The reflective reader will also notice that using this principle we have that *if X is a Banach space and every subspace of X having a Schauder basis has the Dunford-Pettis property, then X has the Dunford-Pettis property; indeed, every subspace of X has the Dunford-Pettis property.*

Since only sequences were used in the proof that every subspace of c_0 has the Dunford-Pettis property, we can conclude to

Theorem 4. *For any set Γ, every closed linear subspace of $c_0(\Gamma)$ has the Dunford-Pettis property.*

A brief detour here might be called for. Theorem 4 cites an intriquing property of c_0. What other Banach spaces share with $c_0(\Gamma)$'s (and their subspaces) the good fortune of possessing the hereditary Dunford-Pettis property? Clearly spaces with the Schur property are so lucky: *every Banach space in which weakly convergent sequences converge in norm is hereditarily Dunford-Pettis*. Beyond these?

After the Schur spaces and $c_0(\Gamma)$ spaces have been disposed of, the next natural candidates might be found among the small C(K) spaces, say for K a dispersed compact Hausdorff space. However, not much comes of looking at these as is seen in the following result due essentially to A. Pelczynski and W. Szlenk [PSz]:

Theorem 5. Let K be a compact Hausdorff space. Then C(K) has the hereditary Dunford-Pettis property if and only if K is dispersed and the ωth derived set of K is empty.

Pelczynski and Szlenk's contribution to Theorem 5 is quite elegant; what they did was find an unconditional normalized basic sequence (e_n) inside $C(\omega^\omega)$ whose coefficient functionals are weakly null. Plainly the closed linear span of the e_n's fails the Dunford-Pettis property. Their construction was based on an earlier construction of J. Schreier [Sch] who displayed a sequence (f_n) of norm one members of C[0, 1] such that even though (f_n) is weakly null no subsequence of (f_n) admits of norm convergent arithmetic means. The connection between such phenomena and the hereditary Dunford-Pettis property is curious; in fact, using a bit of tender love and care, Schreier's construction can be performed in $C(\omega^\omega)$ (see N. Farnum [FBS] or for an alternative but stunningly simple construction J. Bourgain [JB1]). Why is this of interest with regards to the hereditary Dunford-Pettis property? Well, largely because of

Theorem 6. If the Banach space X has the hereditary Dunford-Pettis property then each weakly convergent sequence in X admits a subsequence whose arithmetic means are norm convergent.

The proof of Theorem 6 relies on a remarkable characterization of c_0's unit vector basis discovered by John Elton [Elθ]. Elton's result goes like this: if (x_n) is a normalized weakly null sequence no subsequence of which is equivalent to the unit vector basis of c_0, then (x_n) has a subsequence (y_n) for which given any subsequence (z_n) of (y_n) and any sequence (α_n) of scalars with $(\alpha_n) \notin c_0$ we have

$$\sup_n \left\| \sum_{k=1}^{n} \alpha_k z_k \right\| = \infty .$$

In addition to helping prove Theorem 6 (and through it Theorem 5), this result of Elton's also is instrumental in proving the following result:

Theorem 7. *Every infinite dimensional Banach space contains a subspace isomorphic to c_0, a subspace isomorphic to ℓ_1 or a subspace that fails the Dunford-Pettis property.*

Therefore, hereditarily Dunford-Pettis spaces are c_0 and/or ℓ_1 rich in the sense that every subspace contains either a c_0 or ℓ_1.

Where does Elton's result enter the foray? Suppose you find yourself in the midst of Elton's hypotheses with a normalized weakly null sequence (x_n) having no subsquences equivalent to c_0's unit vector basis. Pass to the horrendous subsequence (y_n). The Bessaga-Pelczynski selection principle allows you to assume (y_n) is basic and, therefore, has an accompanying sequence (y_n^*) of coefficient functionals. Denote by $[y_n]$ the closed linear span of the y_n's and consider the bounded linear projections P_k: $[y_n] \to [y_n]$ defined by $P_k(\Sigma_n y_n^*(x)y_n) = \sum_{n=1}^{k} y_n^*(x)y_n$. Because (y_n) is a basis for $[y_n]$, the P_k's are uniformly bounded in norm. What's more if $x^{**} \in [y_n]^{**}$ then it's easily checked that $P_k^{**}x^{**}$ is just $\sum_{n=1}^{k} x^{**}(y_n^*)y_n$. Therefore, for any $x^{**} \in [y_n]^{**}$ we have

$$\begin{aligned} \sup_k \left\| \sum_{n=1}^{k} x^{**}(y_n^*)y_n \right\| &\leq \sup_k \|P_k^{**}x^{**}\| \\ &\leq \sup_k \|P_k\| \cdot \|x^{**}\| < \infty; \end{aligned}$$

Elton's result forces us to conclude that for each x^{**} in $[y]^{**}$ the sequence $(x^{**}y_n^*)$ belongs to c_0 and so (y_n^*) is weakly null. Since $y_n^*y_n = 1$ for all n it follows that we've shown that *whenever (x_n) is a normalized weakly null basic sequence no subsequence of which is equivalent to the unit vector basis of c_0 then the closed linear span $[y_n]$ of the y_n's fails the Dunford-Pettis property for some subsequence (y_n) of (x_n).*

To prove Theorem F from this it suffices to notice that Rosenthal's ℓ_1-theorem gives the existence of a normalized weakly null basic sequence in any infinite dimensional Banach space that contains no copy of ℓ_1. How? Start with a bounded sequence (x_n) for which $\|x_m - x_n\| \geq \frac{1}{2}$ for $m \neq n$, use Rosenthal's ℓ_1-theorem to extract a weakly Cauchy subsequence (y_n) and look at $z_n = \frac{y_{n+1} - y_n}{\|y_{n+1} - y_n\|}$. If necessary a quick recourse to the Bessage-Pelczynski theorem will produce the desired normalized weakly null basic sequence.

Theorem 6? If (x_n) is a weakly null sequence none of whose subsequences have norm convergent arithmetic means, then (x_n) has a subsequence (y_n) that's bounded away from zero (so might as well be normalized), is basic (if necessary hit the sequence with the Bessaga-Pelczynski principle again) and clearly cannot contain any subsequence equivalent to the unit vector basis of c_0 since careful computation of the arithmetic means of the unit vector basis of c_0 produces a norm null sequence.

Now for Theorem 5.

As we mentioned earlier, the space $C(\omega^\omega)$ contains a weakly null sequence no subsequence of which has norm convergent arithmetic means. If K is a dispersed compact Hausdorff space with non-empty ωth derived set, then there is a continuous map φ of K onto ω^ω; this elegant result is due to J. W. Baker [JWB]. It follows that C(K) contains an isometric copy of $C(\omega^\omega)$ and so C(K) has a subspace failing the Dunford-Pettis property thanks to Theorem 6.

If K is a compact Hausdorff space that contains a non-void perfect subset then going down the checklist of the Main Result of A. Pelczynski and Z. Semadeni [PS], we'll soon find that C(K) contains an isomorph of ℓ_2 and, therefore, is not hereditarily Dunford-Pettis.

In tandem the above two paragraphs show that the only possibilities for hereditarily Dunford-Pettis spaces among the C(K) spaces are those coming from K's that are dispersed and have vanishing ωth derived sets; so let's look at such a C(K). Suppose (f_n) is a basic sequence in C(K). Define the equivalence relationship "$\sim$" on K by $k \sim k'$ whenever $f_n(k) = f_n(k')$ holds for all n. The resulting quotient space $\tilde{K}$ is a dispersed compact metric space whose ωth derived set vanishes; moreover each f_n clearly respects "$\sim$" and so determines a unique $\tilde{f}_n \in C(\tilde{K})$ for which

$f_n(k) \equiv \tilde{f}_n(\pi(k))$ where π: $K \to \tilde{K}$ is the quotient map. It follows that any f in the closed linear span of the f_n's also "lifts" to a unique $\tilde{f} \in C(\tilde{K})$; the mapping $f \to \tilde{f}$ from the closed linear span of the f_n's into $C(\tilde{K})$ is an isometric imbedding! Now an old theorem of S. Mazurkiewicz and W. Sierpinski [MS] tells us that $\tilde{K}$ is homeomorphic to an ordinal space α where the ordinal $\alpha < \omega^{\omega}$; but C. Bessaga and A. Pelczynski [BP2] have shown that such a $C(\alpha)$ must be isomorphic to $C(\omega) = c$ itself isomorphic to c_0, a known hereditarily Dunford-Pettis space. So the closed linear span of the f_n's is isomorphic to a subspace of c_0 and as such has the Dunford-Pettis property. It follows that if K is dispersed and $K^{(\omega)} = \emptyset$, every subspace of C(K) having a Schauder basis has the Dunford-Pettis property and so C(K) is hereditarily Dunford-Pettis

Question 2: Does there exist a hereditarily Dunford-Pettis space X with separable dual that is universal for this class of spaces?

Returning from our detour, we note the following consequence drawn by Grothendieck [GTVS] from the original Dunford-Pettis theorem: *if μ is a regular Borel measure defined on a weakly compact subset of a Banach space X then μ's support is norm separable.* In fact, letting K be the weakly compact set we can define the linear operator T: $L_1(K, \mu) \to X$ by $Tf = \text{Bochner-}\int_K xf(x)d\mu(x)$. It is easy to establish that T takes the closed unit ball of $L_1(K, \mu)$ into the absolutely closed convex hull of K (the computation is virtually identical to one on the top of page 72 of [DU]). It follows that T is a weakly compact linear operator on $L_1(K, \mu)$. But $L_2(K, \mu)$ is dense in $L_1(K, \mu)$ and the closed unit ball of $L_2(K, \mu)$ is weakly compact so by the Dunford-Pettis theorem, T takes the ball of $L_2(K, \mu)$ into a compact subset of X. It follows that T has a norm separable range. We need only notice now that if x_0 is in the support of μ, then x_0 is in the (weak) closure of the range of T. Indeed, each weak neighborhood V of x_0 in K has non-zero μ measure so $\mu(V)^{-1}\chi_V$ belongs to $L_1(K, \mu)$ and $\int_K \mu(V)^{-1}\chi_V \, d\mu = 1$. Moreover, $T(\mu(V)^{-1}\chi_Y)$ is just $\int_K x\chi_V(x)\,\mu(V)^{-1}\,d\mu(x)$ a point in V's weakly closed convex hull. Picking V very small we get members $T(\mu(V)^{-1}\chi_V)$ of T's range very close (in the weak topology) to x_0. This is as it should be.

In response to inquiries from several mathematicians regarding the validity of the similar statement for subspaces of $L_1[0, 2\pi]$, Grothendieck used the fact of C(K) spaces enjoying the Dunford-Pettis property to derive that *if μ is a finite measure and $1 \leq p < \infty$, then any closed linear subspace H of $L_p(\mu)$ consisting entirely of members of $L_\infty(\mu)$ is finite dimensional.* The proof is set forth on p. 178 of [DU]; however, it is so illustrative of Grothendieck's viewpoint (and so pretty!) that it's worth repeating. To start with, we'll denote by H_q, H with the relative $L_q(\mu)$ topology. This in mind we assert that if $p \leq q \leq \infty$ then the linear topological spaces H_q are all isomorphic. To see this notice that H is closed in $L_\infty(\mu)$; further, since for any $p < \infty$, if (f_n) is a sequence in $L_p(\mu)$ converging in p-mean to f then (f_n) admits a subsequence converging to f almost everywhere, it follows that the formal identity map of H_p onto H_∞ has a closed graph. The isomorphic nature of the H_q's ($p \leq q \leq \infty$) follows from this. Now for the punch-line: the inclusion map of $L_\infty(\mu)$ into $L_p(\mu)$ is weakly compact; for $1 < p$, this follows from the reflexivity of $L_p(\mu)$, while for $p = 1$, it follows from the fact that in going from $L_\infty(\mu)$ to $L_1(\mu)$ the inclusion map passes through $L_2(\mu)$. In any case, $L_\infty(\mu) \rightarrow L_p(\mu)$ is a weakly compact linear operator on the C(K)-space(!) $L_\infty(\mu)$. By the Dunford-Pettis property for C(K)-spaces this inclusion map takes weakly compact sets in $L_\infty(\mu)$ onto compact sets in $L_p(\mu)$. But H_∞ is isomorphic to all the H_q's, many of which are, as closed subspaces of reflexive spaces, clearly reflexive; so H_∞'s unit ball is weakly compact and as such is carried by the inclusion map onto a compact set in H_p which is simultaneously a neighborhood of 0 in H_p. Case closed.

This use of the special properties of the natural inclusion

operator is characteristic of the Grothendieck program. To see it in another guise we introduce the notion of a q-absolutely summing operator. An operator $T: X \to Y$ is q-absolutely summing (here $1 \le q < \infty$) if there is a $k > 0$ such that for any $x_1, \cdots, x_n \in X$ we have

$$\sum_{j=1}^{n} \|Tx_j\|^q \le k^q \sup \{\sum_{j=1}^{n} |x^*x_j|^q: \ x^* \in B_{X^*}\}.$$

Let $T: X \to Y$ be q-absolutely summing. Denote by K the closed unit ball B_{X^*} of X^* in its weak* topology. Let π_q be the smallest constant $k > 0$ for which the defining inequality of q-absolutely summing operators holds for T. For any $x_1, \cdots, x_n \in X$ define $\varphi_{x_1,\cdots,x_n} \in C(K)$ by

$$\varphi_{x_1,\cdots,x_n}(x^*) = \pi_q^q \sum_{i=1}^{n} |x^*x_i|^q - \sum_{i=1}^{n} \|Tx_i\|^q;$$

let $\Phi = \{\varphi_{x_1,\cdots,x_n} \in C(K): \ x_1, \cdots, x_n \in X\}$. Notice that since T is q-absolutely summing and π_q is what it is, $\Phi \cap \mathfrak{N} = \emptyset$, where

$$\mathfrak{N} = \{f \in C(K): \ f(x^*) < 0 \text{ all } x^* \in K\}.$$

Moreover, Φ and $\mathfrak{N}$ are convex cones and $\mathfrak{N}$ has an interior. Therefore, there exists a norm-one functional μ on C(K) such that

$$\mu(x) \le 0 \le \mu(\varphi)$$

for all $x \in \mathfrak{N}$ and all $\varphi \in \Phi$. That μ is non-positive on $\mathfrak{N}$ translates to μ being a non-negative measure; μ's being of norm-one says μ is a probability measure on the Borel sets of K; finally, μ's non-negativity on Φ says, in particular, that for any $x \in X$

$$\begin{aligned} 0 &\le \int_K \varphi_x(x^*) d\,\mu(x^*) \\ &= \int_K \{\pi_q^q\, |x^*(x)|^q - \|Tx\|^q\} d\,\mu(x^*) \\ &= \pi_q^q \int_K |x^*(x)|^q \,d\,\mu(x^*) - \|Tx\|^q \end{aligned}$$

since μ is a probability measure. Of course, the result is that *if $T: X \to Y$ is q-absolutely summing, then there is a probability Borel measure μ on (B_{X^*}, weak^*) such that for any $x \in X$*

$$\|Tx\|^q \leq \pi_q^q \int |x^*(x)|^q d\,\mu(x^*) \ .$$

This result is sometimes known as the Grothendieck-Pietsch Domination Theorem.

So what? Well a bit of reflection shows that the above inequality can be interpreted in such a fashion that any q-absolutely summing operator $T: X \to Y$ admits a factorization of the form

$$\begin{array}{ccc} X & \xrightarrow{T} & Y \\ A & & B \\ & S_q & \end{array}$$

where S_q is a subspace of an $L_q(\mu)$ and $A: X \to S_q$ and $B: S_q \to Y$ are bounded linear operators. What's more, the operator $A: X \to S_q$ which takes $x \in X$ to $x(\cdot) \in S_q$ actually factors through the $C(K)$-space $L_\infty(\mu)$. It follows that *every q-absolutely summing operator is weakly compact and completely continuous; of course, a quick consequence of this is that given any two summing operators*--be one 11-absolutely summing and the other 37.3-absolutely summing--*whatever each individual operator's degree of absolute summability--the composition is compact.*

Though the above is not a true application of the Dunford-Pettis property for $C(K)$ spaces, the discussion of q-absolutely summing operators does lead to such. The case in mind is $q = 1$; in this case, we refer to the operators simply as absolutely summing. We've seen that every q-absolutely summing operator factors rather canonically through some Dunford-Pettis space (a $C(K)$ in fact); for absolutely summing operators there's another possibility: an $L_1(\mu)$. Now every q-absolutely summing operator factors through a subspace of $L_q(\mu)$. It was

Grothendieck [GRes] who asked if every absolutely summing operator factors through the whole of an $L_1(\mu)$ space. For our present purposes this possibility is intriquing on at least three counts: first, L_1 spaces have the Dunford-Pettis property; second, the dual of an L_1-space is an L_∞-space and operators on L_∞ spaces are well understood and for the most part weakly compact--more on this later; third, L_1-spaces are weakly complete and their weakly compact subsets well understood. Unfortunately, not every 1-absolutely summing operator factors through an L_1-space; in their Acta paper [GLust], Y. Gordon and D. R. Lewis showed that there are many spaces of operators that act as domains of natural absolutely summing operators which cannot be so factored. Gordon and Lewis go on to characterize those spaces X for which good factorizations of 1-absolutely summing operators exist. To describe their result we need a bit of terminology: the *unconditional basis constant* u(X) of a Banach space X is the least constant λ having the property that there exists a Schauder basis (x_n) for X with

$$\|\Sigma_n \pm x_m^*(x)x_n\| \le \lambda$$

for any vector $\Sigma\, x_n^*(x)x_n$ in X of norm one and any choices of signs $\pm$. The *local unconditional constant* lust (X) of X is the infimum of those λ's with the following property: given a finite-dimensional subspace Y of X there exists a space U and operators A: $Y \to U$ and B: $U \to X$ such that $BA|_F = id_F$ and $\|A\|\,\|B\|u(U) \le \lambda$. The Gordon-Lewis result: *In order that a Banach space X have finite lust constant it is necessary and sufficient that every 1-absolutely summing operator T: X → Y admit a factorization of the form*

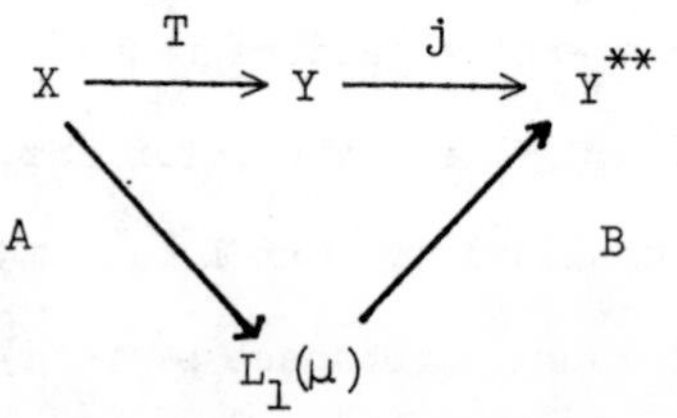

where j is the natural imbedding of Y into Y^{**}, μ is some finite measure and A, B are bounded linear operators.

It is noteworthy that among those spaces X which satisfy lust (X) $< \infty$ one finds all the $\mathcal{L}_p$-spaces ($1 \leq p \leq \infty$) of J. Lindenstrauss and A. Pelczynski [LP] and all complemented subspaces of spaces having unconditional bases; this latter fact is noted by Gordon and Lewis while the former is proved in the paper [DPR] of E. Dubinsky, A. Pelczynski and H. P. Rosenthal. As an example of how this factorizability of absolutely summing operators can be used in tandem with the Dunford-Pettis theorem we take the opportunity to repeat the beautiful proof due to A. Pelczynski that *the disk algebra A, though it has a Schauder basis, does not possess an unconditional basis.* In presenting this proof, we shall have need of a special result regarding $L_\infty(\mu)$-spaces that will be worth revisiting on completing our discussion of the Pelczynski result.

Suppose the disk algebra A has an unconditional basis. By the Gordon-Lewis result, it'd follow that every absolutely summing operator T: A $\to$ Y on being followed by the natural inclusion of Y into Y^{**} can be factorized through an $L_1(\mu)$-space where μ is some finite measure. Note that if Y were isomorphic to a dual space then the need for passing up to Y^{**} for this factorization would not exist since Y would be complemented in Y^{**} in this case. Let's look at the special absolutely summing operator i: $A \to H^1$ of inclusion of A into the Hardy space H^1. H^1 is a known dual and so if A has an unconditional basis we'd have that i admits a factorization

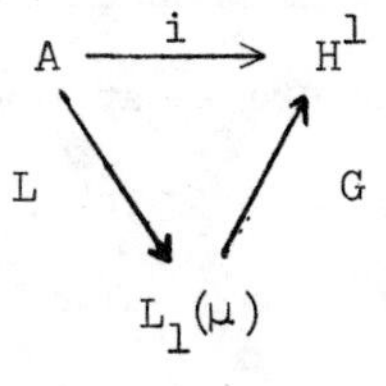

for some bounded linear operators G, L and some $L_1(\mu)$ space with μ finite. Observe that the operator G: $L_1(\mu) \to H^1$ is completely continuous by the

Dunford-Pettis theorem (and this is much of the justification for the inclusion of this result in these notes: it at least enters into the spirit of the Dunford-Pettis property), since H^1 is a separable dual. Let's look at the operator $L: A \to L_1(\mu)$. Its adjoint $L^*: L_\infty(\mu) \to A^*$ is not just your average operator. In fact we know that (in the language of duality) the brothers Riesz showed that A^* is the ℓ_1 direct sum of a separable Banach space L_1/H_0^1 and the weakly sequentially complete space S of finite Borel measures on the circle singular with respect to Lebesgue measure. Neither component of A^* contains a copy of ℓ_∞ so their sum also has no such copy. Here's the point: any operator from $L_\infty(\mu)$ to a Banach space which doesn't contain a copy of ℓ_∞ is weakly compact; this result is due to H. P. Rosenthal [HPRL] and provides another contact of the Dunford-Pettis property with the basic structure of the classical Banach spaces, a contact we will return to momentarily. So L^* (and hence L) is weakly compact. But now as we know that G is completely continuous and L is weakly compact we can deduce that $GL = i: A \to H^1$ is compact. This is easily seen to be false, however, providing us with our contradiction.

In the above proof we encountered an extremely important phenomenon that cannot be divorced from the study of the Dunford-Pettis property. Frequently a Banach space of special character makes it easy for an operator defined on it to be weakly compact. When this occurs in a space with the Dunford-Pettis property some surprisingly strong conclusions may be drawn. We take this opportunity to dip into this grab-bag of Banach spaces on which it is easy for an operator to be weakly compact and discuss briefly two special classes of Banach spaces which arise in this connection and whose general structure is at least as mystifying as that of spaces enjoying the Dunford-Pettis property.

To introduce the first class of spaces we need to say a word about series in Banach spaces. If $\Sigma_n x_n$ is a formal series in X we say $\Sigma_n x_n$ is weakly unconditionally Cauchy (in short $\Sigma_n x_n$ is a wuc) if for each $x^* \in X^*$, $\Sigma_n|x^*x_n| < \infty$. Following A. Pelczynski [PV] we say that X enjoys the property V (for Vitali?) if any subset K of X^* for which

$$\lim_n x^* x_n = 0 \qquad \text{uniformly for } x^* \in K$$

for each wuc $\Sigma_n x_n$ in X is relatively weakly compact.

If an operator T: X → Y transforms wucs into unconditionally convergent series then we call T an unconditionally converging operator; it is a frequently useful and certainly elegant result due to C. Bessaga and A. Pelczynski [BP1] that an operator T: X → Y is unconditionally converging if and only if there is no isomorph of c_0 inside X on which T is an isomorphism; oftentimes we just say that T is unconditionally converging if and only if T fixes no copy of c_0. Pelczynski's property V? A Banach space X has property V if and only if every unconditionally converging operator T: X → Y is weakly compact. This result is due to Pelczynski [PV]; its proof is simple enough to deserve presentation herein.

Suppose X has property V and T: X → Y is an unconditionally converging operator. We'll show that T^* is weakly compact. Let (y_m^*) be a bounded sequence in Y^* and $\Sigma_n x_n$ be a wuc in X. Then $\Sigma_n Tx_n$ is an unconditionally convergent series in Y and so in particular $\|Tx_n\| \to 0$. Thus (Tx_n) tends to zero uniformly on bounded subsets of Y^*, hence

$$\begin{aligned} 0 &= \lim_n (Tx_n)(y_m^*) \\ &= \lim_n (T^* y_m^*)(x_n) \qquad \text{uniformly in m.} \\ &= \lim_n x_n(T^* y_m^*) \end{aligned}$$

Since X has property V, it follows that $\{T^* y_m^*\}$ is relatively weakly compact.

In the opposite direction we notice that if you start with a wuc $\Sigma_n x_n$ then given any sequence (F_n) of finite pairwise disjoint sets of natural numbers the series $\Sigma_n(\Sigma_{k\in F_n} x_k)$ is also a wuc. Moreover, the only thing that could keep a wuc $\Sigma_n x_n$ from being unconditionally convergent would be the failure of $(\|\Sigma_{k\in F_n} x_k\|)$ to tend to zero for some such sequence (F_n).

This in mind suppose $K \subseteq X^*$ satisfies

$$\lim_n \sup_{x^* \in K} |x_n(x^*)| = 0$$

for any wuc $\Sigma_n x_n$ in X. Define the operator T: $X \to \ell_\infty(K)$ by $Tx = (x^*x)_{x^* \in K}$. By the opening comments of this paragraph we see that T is an unconditionally converging operator hence is weakly compact. It follows that T^* is also weakly compact. But clearly K is contained in $\{T^*\delta_k : k \in K\}$ where $\delta_k \in \ell_\infty(K)^*$ is given by evaluation on the kth coordinate. So K is relatively weakly compact as was to be proved.

The interest in Pelczynski's result herein is found in the fundamental fact demonstrated by him that C(K) spaces enjoy property V. It follows that whenever X is a Banach space containing no copy of c_0 every operator T: $C(K) \to X$ is weakly compact. But all C(K) spaces also have the Dunford-Pettis property so that any complemented infinite dimensional subspace Y of a C(K) contains a copy of c_0. Why is this? Well notice that if Y is a complemented subspace of C(K) having no copy of c_0 inside it then any continuous linear projection P of C(K) onto Y is weakly compact by Pelczynski's result. However a basic fact about spaces with the Dunford-Pettis property (and possibly characteristic of such spaces) is the following: if X has the Dunford-Pettis property and T: $X \to X$ is a weakly compact linear operator then T^2 is compact. It follows that the projection P has a compact square $P^2 = P$ and so P is itself a compact operator. From this we see that Y must be finite dimensional.

Noteworthy here is the use of the possession by C(K) spaces of both the Dunford-Pettis property and property V. Many other C(K)-like function spaces enjoy both properties as well; for instance, the disk algebra A has both properties (we'll talk a little bit more about the Dunford-Pettis property for A in a moment) as do many uniform algebras (see [PKSU] for a discussion of this). In each such case the conclusion that complemented infinite dimensional subspaces contain copies of c_0 holds true. The pair Dunford-Pettis and V is indeed a dynamic duo.

A second class of Banach spaces on each member of which it is easy to find many weakly compact operators was introduced by Grothendieck in [GDP]; these spaces have come to be known as Grothendieck spaces. A Banach space X is called a Grothendieck space if weak* null sequences in X^* are weakly null. If Y is in some sense a "small" Banach space then every operator from a Grothendieck space to Y is weakly compact. Here "small" can mean separable, weakly compactly generated, separably complementable (i.e., separable subspaces are contained in separable complemented subspaces), having weak* sequentially compact dual balls, or even failure to contain a copy of ℓ_1. Indeed, if B_{Y^*} is weak* sequentially compact or if Y fails to contain a copy of ℓ_1 then any operator on a Grothendieck space domain with values in Y is weakly compact. Among those spaces which are known to be Grothendieck spaces one finds many "large" C(K) spaces. In particular, if K is extremally disconnected (= Stonian) or basically disconnected (= σ-Stonian) or even just an F-space (in the sense of L. Gillman and M. Jerison [GJ]) then C(K) is a Grothendieck space; these facts are due respectively to Grothendieck [GDP], T. Ando [A] and G. Seever [GS]. On the other hand, the classical Baire spaces are now known to be Grothendieck spaces as are a number of other C(K) spaces that suitably mimic the Baire classes in their order structure; for more on this we refer the reader to F. Dashiell [D]. In most of these cases one can strengthen the statements regarding when an operator on a Grothendieck space is weakly compact considerably. For instance, Rosenthal has shown in [HRL] that if K is σ-Stonian and T: $C(K) \to X$ fixes no copy of ℓ_∞ then T is weakly compact. It follows as above then that for σ-Stonian K's, infinite dimensional complemented subspaces of C(K) contain copies of ℓ_∞. For general Grothendieck spaces the best that is known seems to be the following consequence of Rosenthal's ℓ_1-theorem found in [DS]: if X is a Grothendieck space and T: $X \to Y$ is an operator that takes bounded sequences to sequences with weakly Cauchy subsequences then T is weakly compact. It follows that an operator T on a Grothendieck space is weakly compact if and only if T fixes no copy of ℓ_1. Curiously, it might well be that all Grothendieck spaces have property V. Such is not known and appears to be nontrivial (if true). It is a part of the folklore of the subject that for dual spaces property V already implies the Grothendieck property. Since this is our last real detour from the main path of this survey we will provide the reader with a quick proof of the folklore result and mention one or two open problems in connection with these two properties.

So suppose X is a Banach space whose dual X^* has property V and let (x_n^{**}) be a weak* null sequence in X^{**}. We'll show that $\{x_n^{**}\}$ is relatively weakly compact in X^{**}. To do this look at a typical wuc $\Sigma_n x_n^*$ in X^* and notice that for any set Δ of natural numbers,

$$x_\Delta^* = \text{weak}^* \lim_{n\to\infty} \Sigma_{i\in\Delta, i\le n} x_i^*$$

exists. Define $\mu_n \in \ell_\infty^*$ by

$$\mu_n(\Delta) = x_n^{**} x_\Delta^*.$$

Since (x_n^{**}) is weak* null, $\lim_n \mu_n(\Delta) = 0$ for each Δ. By Phillips' Lemma [DU, p. 33]

$$0 = \lim_n \Sigma_{i=1}^\infty |\mu_n(\{i\})| = \lim_n \Sigma_{i=1}^\infty |x_n^{**} x_i^*|.$$

Property V now comes into the play to give $\{x_n^{**}\}$ relatively weakly compact and end the game.

As we mentioned the following is open.

Question 3. Do Grothendieck spaces have property V?

Property V's dependence on the weak compactness of unconditionally converging operators makes the next question interesting. I believe an affirmative solution to this question will be difficult (perhaps because such is not forthcoming):

Question 4. If T: $X \to Y$ is an unconditionally converging operator need T factor through a space P that contains no copy of c_0?

One conclusion to be drawn from an affirmative answer to the above would be that all C^* algebras have property V. In particular, $\mathcal{B}(H)$ would be a Grothendieck space.

Enough for our detour.

Returning to the discussion of spaces with the Dunford-Pettis property, if only $C(K)$'s and $L_1(\mu)$'s among the naturally encountered spaces have the Dunford-Pettis property, it is indeed a very special property. However, many other spaces enjoy this property. To see this we start with a modest improvement of a result of S. Kisliakov [Kis 1] which was obviously suggested by his result. First, a result from R. H. Lohman [RHL], which itself depends on the Rosenthal ℓ_1 theorem [Rℓ_1]. Recall the statement of this gem (or rather that of the Rosenthal-Dor ℓ_1 theorem): *any bounded sequence (x_n) in a Banach space having no weakly Cauchy subsequence admits of a subsequence (y_n) that's equivalent to ℓ_1's unit vector basis.*

Lemma 8. *Let Y be a closed linear subspace of X and suppose that $Y \not\supseteq \ell_1$. Then each weakly Cauchy sequence in X/Y has a subsequence that's the image of a weakly Cauchy sequence in X under the natural quotient map $q: X \to X/Y$.*

Proof. Suppose (q_n) is a weakly Cauchy sequence in X/Y no subsequence of which comes from a weakly Cauchy sequence in X. Using the Rosenthal-Dor-ℓ_1-theorem, we can find a ℓ_1-unit vector basis (x_n) in X that's mapped onto a subsequence of (q_n) which for convenience we'll call (q_n). The sequence $(q_{2n} - q_{2n-1})$ is weakly null so there's a strictly increasing sequence (k_n) of positive integers and a sequence (σ_n) in X/Y tending to zero in norm with each σ_n a convex combination of members of $\{q_{2j} - q_{2j-1}: k_n \leq j < k_{n+1}\}$. If (e_n) is the corresponding sequence of convex combinations of the $x_{2j} - x_{2j-1}$'s, then (e_n) is a ℓ_1-unit vector basis in X. Of course, q takes e_n to σ_n. But distance $(e_n, Y) \to 0$ because of this. Hence Y contains a sequence that is close enough to a suitable subsequence of (e_n) to insure its equivalence to that subsequence; since subsequences of the unit vector basis of ℓ_1 span copies of ℓ_1, we've found that Y contains an isomorph of ℓ_1.

Note: It is an easy consequence of Lohman's Lemma and the Rosenthal-Dor-ℓ_1-theorem that the non-containment of ℓ_1 is a "three-space property," i.e., *TFAE: 1) X contains no copy of ℓ_1; 2) for each subspace Y of X neither Y nor X/Y contains a copy of ℓ_1; 3) there is a subspace Y of X such that neither Y nor X/Y contains a copy of ℓ_1.*

Theorem 9. *Suppose X has the Dunford-Pettis property and Y is a closed linear subspace of X with $Y \not\supseteq \ell_1$. Then X/Y has the Dunford-Pettis property.*

Proof. Let (q_n) and (q_n^*) be weakly null sequences in X/Y and $(X/Y)^*$ respectively and take a strictly increasing sequence (k_n) of positive integers. By the previous lemma there is a subsequence (j_n) of (k_n) such that $q_{j_n} = qx_n$ for some weakly Cauchy sequence (x_n) in X, where $q\colon X \to X/Y$ denotes the natural quotient map. Now (x_n) is weakly Cauchy in X and $(q_{j_n}^*)$ is weakly null in $(X/Y)^*$ hence in X^*. Since X has the Dunford-Pettis property, $\lim_n q_{j_n}^*(x_n) = 0$. But $q_{j_n}^*(x_n) = q_{j_n}^*(q_{j_n})$ and we've shown that $\lim_n q_n^* q_n = 0$.

Corollary 10 (S. Kisliakov). *If X is a reflexive subspace of $L_1(\mu)$ then $L_1(\mu)/X$ has the Dunford-Pettis property.*

It ought to be remarked here that M. Kadec and A. Pelczynski [KP] have demonstrated that the reflexivity of a subspace X of $L_1(\mu)$ is entirely equivalent to the failure of X to contain a copy of ℓ_1. On the other hand, in spaces like C[0, 1] there are lots of "natural" spaces containing no ℓ_1's that are not reflexive.

If one uses a bit of duality then the following might be the result:

Corollary 11. *Suppose X^* has the Dunford-Pettis property and Y is a subspace of X for which $Y^\perp$ contains no isomorph of ℓ_1 then Y^* has the Dunford-Pettis property.*

Indeed, Y^* is isomorphic to $X^*/Y^\perp$ which has the Dunford-Pettis property through the wise choice of hypotheses and the good graces of Theorem 9.

Returning to Kisliakov's theorem, it is of interest to know for which subspaces X of L_1 does L_1/X have the Dunford-Pettis property. Indeed, it is plausible that the following have affirmative answers:

Question 5. If X is isomorpic to a dual space and X imbeds in $L_1(\mu)$, need $L_1(\mu)/X$ have the Dunford-Pettis property?

More specifically (but probably not equivalent)

Question 5'. If X is (isomorphic to) a dual subspace of $L_1[0, 1]$, then need $L_1[0, 1]/X$ have the Dunford-Pettis property?

An important case wherein the answer is known is $X = H_0^1 \subseteq L_1(T)$, where T is the unit circle with normalized Lebesgue measure and H_0^1 is the collection of complex-valued functions f analytic on the open disc with boundary values in $L_1(T)$ for which $f(0) = 0$. In this case we have the

Theorem 12 (J. Chaumat). *$L_1(T)/H_0^1$ has the Dunford-Pettis property.*

The importance of Theorem 8 derives in large part to the fact that it has as a consequence

Corollary 13 (J. Chaumat). *The disk algebra A has the Dunford-Pettis property.*

In fact, the classical F and M. Riesz Theorem tells us (in the language of duality) that A^* is identifiable with $(L_1/H_0^1) \oplus S$ where S is the space of regular Borel measures on T that are singular with respect to Lebesgue measure. Since S is a complemented subspace of $C(T)^*$, S has the Dunford-Pettis property; since L_1/H_0^1 has this property too, the direct sum, $(L_1/H_0^1) \oplus S$, is easily seen to have it. So A^* has the Dunford-Pettis property and hence so too does A.

Now the proof of Theorem 12 is quite a bit deeper than the proof of the corresponding fact about L_1 and in particular depends on that result, along with a lifting theorem for weakly compact sets. An excellent source from which to learn the proof and related developments is §§6-8 of Pelczynski's monograph [PKSU].

In passing it is to be noted that Theorem 12's proof depends very strongly on the position of H_0^1 in $L_1(T)$ in that H_0^1 is weak* closed in $C(T)^*$. From this it follows that from each coset of $L_1(T)/H_0^1$ one can pick a member of the coset whose L_1-norm is exactly the coset norm. In particular, we must admit that there is at present somewhat scant evidence towards an affirmative response to Questions 2 and 2'. A bit more evidence can be gleamed from the fact that for all the known isomorphs Y of ℓ_1 found in $L_1[0, 1]$, L_1/Y is isomorphic to L_1 and so has the Dunford-Pettis property. It may well be (and certainly is if Question 3 has an affirmative response!) that L_1/Y always has the Dunford-Pettis property for isomorphs Y of ℓ_1; I don't know even this. It should be mentioned, however, that it is a long-standing open problem whether or not whenever an isomorph of ℓ_1 appears in L_1 it is complemented; it is a consequence of the Enflo-Starbird theorem (cf. [DU], pp. 94-95 for a brief discussion of this result) that any complement of an ℓ_1 in $L_1[0, 1]$ must be isomorphic to $L_1[0, 1]$.

A word to the wise here. After it is known that C(K)'s and L_1's have the Dunford-Pettis property, it is only too natural to try to prove other classical (and neo-classical) spaces patterned somewhat after these spaces also have the property, especially if these latter spaces have similarly defined norms. Sometimes (as in the case of the disk algebra) this effort is rewarded; oftentimes, it is not. I know of no case where the reward (when it comes) is easily attained. Let me mention a few places where negative results are known and a few places where the enjoyment of the Dunford-Pettis property is, as yet, unknown.

First, a couple of negative results.

H^1 fails the Dunford-Pettis property. Indeed, H^1 is the dual of the space VMO of functions of vanishing mean oscillation (cf. [DS]). By H^1's separability, VMO cannot contain an isomorph of ℓ_1. Moreover, VMO fails the Dunford-Pettis property as is easily seen through Theorem 3; were VMO to have the property, H^1 would have the Schur property which it doesn't contain, as it does, subspaces (complemented ones even) isomorphic to Hilbert space (a good reference on the Banach space view of this example is S. Kwapien and A. Pelczynski [KWP]).

The Lorentz spaces $\Lambda(w, 1)$ $[L_0]$ fail the Dunford-Pettis property. The argument is modelled on the previous example. Key are the facts that all $\Lambda(w, 1)$'s are isomorphic to duals and have the Radon-Nikodym property. It follows that whatever their preduals might be these preduals don't contain ℓ_1 nor do they have the Dunford-Pettis property. This latter because if they did, again by Theorem 3, $\Lambda(w, 1)$ would be a Schur space. But each $\Lambda(w, 1)$ contains an isomorph of Hilbert space, namely, the span of the Rademacher functions (or variations of them) in $\Lambda(w, 1)$ is isomorphic to Hilbert space. A good source for these Banach space properties of $\Lambda(w, 1)$ spaces is the Memoir (MON] of W. B. Johnson, B. Maurey, L. Tzafriri (who was the first to point out to me the fact that the $\Lambda(w, 1)$'s all have the Radon-Nikodym property) and G. Schechtman.

Some open ends are included in the following questions.

Question 6. For $n \geq 2$ and $k \geq 1$ does the space $C^k(K^n)$ of all scalar-valued functions f defined on $I^n = [0, 1]^n$ that have continuous partial derivatives of all orders $\leq k$ have the Dunford-Pettis property?

Here we norm $f \in C^k(I^n)$ by

$$\|f\| = \sup_{|\alpha| \leq k} \left\| \frac{\partial^\alpha f}{\partial^{\alpha_1} x_1 \cdots \partial^{\alpha_n} x_n} \right\|_\infty$$

where $\alpha = (\alpha_1, \cdots, \alpha_n)$ is a multi-index and $\frac{\partial^\alpha f}{\partial^{\alpha_1} x_1 \cdots \partial^{\alpha_n} x_n}$ has the standard meaning. It might be noted that it is known that if $n \geq 2$ then $C^k(I^n)$ is not isomorphic to a complemented subspace of any $C(K)$ space, a result due to S. Kisliakov [Kis 2].

Question 7. Which uniform algebras have the Dunford-Pettis property?

F. Delbaen [FD] has the most incisive result pertaining to this problem; Delbaen's result: if U is a uniform algebra modelled on the compact space K (so U is a closed subalgebra of complex C(K) that contains the constants and separates points of K) for which the collection of regular Borel measures on K representing any given complex homomorphism on U is relatively weakly compact, then U has the Dunford-Pettis property.

Question 8: For which compact convex sets K (in locally convex spaces) does the space A(K) of continuous real-valued affine functions defined on K have the Dunford-Pettis property?

It is known that whenever K is a Choquet simplex then A(K)'s dual is an L_1 space; see A. Lazar and J. Lindenstrauss [LLL_1] for a proof of this. In particular, whenever K is a Choquet simplex, A(C) has the Dunford-Pettis property. This does not tell the whole story however. Paula Saab [PSθ] has observed that if X is a real Banach space with the Dunford-Pettis property and K is B_{X^*} in the weak* topology, then A(K) has the Dunford-Pettis property. Indeed, X is isomorphic to a subspace of codimension 1 in A(K). Paula has gone on to notice that if K_1 is a Choquet simplex and K_2 is the dual ball (in its weak* topology) of a Banach space having the Dunford-Pettis property then $K = K_1 \times K_2$ is a compact convex set, of a different sort than either K_1 or K_2, for which A(K) has the Dunford-Pettis property.

We close the discussion of spaces for which the enjoyment of the Dunford-Pettis property remains open by considering spaces of vector-valued functions. Here the information is fragmentary and the natural open problems many. To pick and choose typical problems might well be best and so we state the following well-known questions.

Question 9. If K is a compact Hausdorff space and X is a Banach space with the Dunford-Pettis property, need the space $C_X(K)$ of continuous X-valued functions defined on K have the Dunford-Pettis property?

Question 10. If (Ω, Σ, μ) is a measure space and X is a Banach space enjoying the Dunford-Pettis property does the space $L_1(\mu, X)$ of Bochner μ-integrable X-valued functions have the Dunford-Pettis property?

These two questions are really quite closely related and it seems likely that an affirmative response to one will lead to an affirmative response to the other. The difficulty in answering either is tied up with determining criteria for weak compactness in the space M(K, X) of regular Borel X-valued measures of bounded variation defined on a compact Hausdorff space K. At present, no general criteria exist and to be honest a reasonable set of criteria may not exist. In any case all bits of partial information regarding Questions 6 and 7 have been hard to come by and each is deserving of some discussion.

It was Grothendieck [GMEM] who first explicitly noticed that

(a) *if K_1 and K_2 are compact Hausdorff spaces then $C_{C(K_2)}(K_1)$ is isometrically isomorphic to $C(K_1 \times K_2)$ and so has the Dunford-Pettis property*; and

(b) *if μ_1, μ_2 are any countably additive measures then $L_1(\mu_1, L_1(\mu_2))$ is isometrically isomorphic to $L_1(\mu_1 \times \mu_2)$ and so has the Dunford-Pettis property as well.*

It took a long while before the next bits of information came in. The first of these bits came from Ivan Dobrakov [DOB] who showed

(c) *if K is a compact Hausdorff space and X has the Schur property then $C_X(K)$ has the Dunford-Pettis property.*

Though many people were at this time trying to get a handle on weakly compact sets in $L_1(\mu, X)$ none were very successful. The least unsuccessful were possibly J. K. Brooks and N. Dinculeanu [BD] who pushed the old Dunford proof through to its last conceivable conclusions and J. Diestel [SIMP] where it was noticed that the Davis-Figiel-Johnson-Pelczynski machinery allowed one to graft the classical Dunford Proof to slightly different surfaces. In either case the resulting conditions for relative weak compactness in $L_1(\mu, X)$ must be seen to be believed and each in far short of being both necessary and sufficient. Not to be discouraged, Kevin Andrews [KA] attached the question of the Dunford-Pettis property for $L_1(\mu, X)$ and contributed several interesting results; the first of which goes as follows:

(d) *if X^* is a Schur space then $L_1(\mu, X)$ has the Dunford-Pettis property;*

To describe the second we need Andrews' notion of a Dunford-Pettis set: a subset K of a Banach space X is called a Dunford-Pettis set if whenever $(x_n) \subseteq K$ and $(x_n^*) \subseteq X^*$ with $x_n^* \to 0$ (weakly) then $x_n^* x_n \to 0$. Andrews showed

(e) *if X has the Dunford-Pettis property and $K \subseteq L_1(\mu, X)$ is a weakly compact set of either sort described in [BD] or [SIMP], then K is a Dunford-Pettis set and so, for any weakly compact operator $T: L_1(\mu,X) \to Y$, TK is relatively compact.*

All the results (a)-(e) above have been obtained by getting a handle on weak compactness in spaces of integrable functions. This brings us to the most recent (and most spectacular) bit of progress regarding questions 6 and 7; J. Bourgain [BDP] has shown

(f) *for any countably additive measure μ and any compact Hausdorff space K, the spaces $L_1(\mu, C(K))$ and $C_{L_1(\mu)}(K)$ have the Dunford-Pettis property.*

Bourgain's proof is exceptionally clever depending in the end on a non-linear trick that is very special to proving the ℓ_∞-sum of L_1 spaces has the Dunford-Pettis property. This non-linear technique is not the only unusual feature of Bourgain's paper however and it seems that some other features of this work are worth discussing in more detail; this is not to say Bourgain's non-linear trick is unworthy of but a mention--it's just too technical to take up space here.

In addition to maneuvering some non-linear aspects into the study of the Dunford-Pettis property (and these are introduced explicitly to circumvent the difficulties of characterizing weakly compact sets in $L_1(C)$ and $C^*_{L_1}$), Bourgain brings finite dimensional methods to the fore in the Dunford-Pettis matters. Key to his overall development is the following elegant

Theorem 14 (Bourgain). *Let X be a Banach space and assume $X = \overline{\cup_n X_n}$, where (X_n) is an increasing sequence of closed linear subspaces of X for which $(\Sigma \oplus X_n)_{\ell_\infty}$ has the Dunford-Pettis property. Then X has the Dunford-Pettis property.*

Proof. Let (x_n) and (x^*_n) be weakly null sequences in X and X^* respectively. Could it be that $x^*_n x_n$ be non-null? Clearly, we may as well assume that each x_n belongs to $\cup_n X_n$. Equally clear is the fact that we can, by passing to a subsequence if necessary, assume that $x_1, \cdots, x_n \in X_n$. In fact, the act of passing to a subsequence of the X_n's doesn't affect the density of their union nor does it affect their ℓ_∞-sum having the Dunford-Pettis property, since the ℓ_∞-sum of a subsequence is complemented in the ℓ_∞-sum of the whole sequence.

Let $i_n\colon X_n \to X$ be the natural injection and $p_n\colon (\Sigma \oplus X_n)_{\ell_\infty} \to X_n$ be the natural projection. Look at the sequence $(\mathfrak{X}_m)$ in $(\Sigma \oplus X_n)_{\ell_\infty}$ given by

$$\mathfrak{X}_1 = (x_1, x_1, x_1, x_1, \cdots)$$

$$\mathfrak{X}_2 = (0, x_2, x_2, x_2, \cdots)$$

$$\mathfrak{X}_3 = (0, 0, x_3, x_3, \cdots)$$

$$\vdots$$

For any m, n, if we look at $p_n^* i_n^*(x_m^*)$ we get a vector in $(\Sigma \oplus X_n)_{\ell_\infty}^*$ whose norm does not exceed $\sup_m \|x_m^*\| < \infty$. Therefore, if we take a free ultrafilter $\mathcal{U}$ on N, for each m

$$\lim_{\mathcal{U}} p_n^* i_n^*(x_m^*) = \mathfrak{x}_m^*$$

exists in $(\Sigma \oplus X_n)_{\ell_\infty}^*$. What's more

$$\mathfrak{x}_m^*(\mathfrak{x}_m) = \lim_{\mathcal{U}} x_m^*(\mathfrak{x}_n(m)) = x_m^*(x_m).$$

If $(\mathfrak{x}_m)$ and $(\mathfrak{x}_m^*)$ are weakly null in their respective spaces, namely in $(\Sigma \oplus X_n)_{\ell_\infty}$ and $(\Sigma \oplus X_n)_{\ell_\infty}^*$, then our hypothesis on $(\Sigma \oplus X_n)_{\ell_\infty}$ will carry us to the conclusion that $\mathfrak{x}_n^* \mathfrak{x}_n \to 0$ and therefore to the completion of the proof.

Fix an infinite subset M of N and a $\delta > 0$.

Because (x_n) is weakly null there is a finite collection $\{\lambda_i\}$ of positive reals indexed over members i of M such that $\Sigma \lambda_i = 1$ and $\|\Sigma_i \lambda_i \epsilon_i x_i\| \leq \delta$ for any choice $\epsilon_i = \pm 1$ of signs. It follows from this that

$$\|\Sigma_i \lambda_i \mathfrak{x}_i\| = \sup_n \|\Sigma_{i \leq n} \lambda_i x_i\| \leq \delta;$$

in fact, if we fix n then

$$\begin{aligned} &\|\Sigma_{i \leq n} \lambda_i x_i\| \\ &\quad = \|\tfrac{1}{2}(\Sigma_{i \leq n} \lambda_i x_i + \Sigma_{\substack{i \in M \\ n<i}} \lambda_i x_i) + \tfrac{1}{2}(\Sigma_{i \leq n} \lambda_i x_i + \Sigma_{\substack{i \in M \\ n<i}} - \lambda_i x_i)\| \\ &\quad \leq \delta. \end{aligned}$$

In particular, $(\mathfrak{x}_n)$ is weakly null in $(\Sigma \oplus X_n)_{\ell_\infty}$.

What about $(\mathfrak{x}_n^*)$? Well (x_n^*) is weakly null and so there's a convex combination $\Sigma_{i \in M} \lambda_i x_i^*$ of norm $\leq \delta$. But then for any n

$$\begin{aligned} \|\Sigma_{i \in M} \lambda_i p_n^* i_n^*(x_i^*)\| &\leq \|p_n^* i_n^*\| \, \|\Sigma_{i \in M} \lambda_i x_i^*\| \\ &\leq \delta \end{aligned}$$

from which it follows that

$$\|\Sigma_{i \in M} \lambda_i \mathfrak{X}_i^*\| \leq \delta$$

and $(\mathfrak{X}_n^*)$ is also weakly null.

Using Theorem 14 in tandem with a bit of the theory of ultraproducts of Banach spaces and his proof that the ℓ_∞-sum of L_1-spaces has the Dunford-Pettis property, Bourgain goes on to prove

Theorem 15 (Bourgain). *Let $\lambda < \infty$ be given. Suppose X has the property that given any finite dimensional subspace U of X there is a subspace V of X containing U whose Banach-Mazur distance from some $(\ell_1^k \oplus \cdots \oplus \ell_1^k)_{\ell_\infty}$ is no more than λ. Then any ultra-product $X_\mathcal{U}$ of X and all the duals of X have the Dunford-Pettis property.*

In particular, for any measure μ and any compact Hausdorff space K the spaces $L_1(\mu, C(K))$, $C_{L_1(\mu)}(K)$ and all their duals have the Dunford-Pettis property.

It is clear that though the information regarding Questions 6 and 7 is fragmentary, it is positive as far as it goes. The Dunford-Pettis property does seem to be respectful of the least and greatest cross norms; indeed encompassing Questions 6 and 7 are the next two.

Question 11. If X and Y have the Dunford-Pettis property, need their injective tensor product $X \check{\otimes} Y$?

Question 12. If X and Y have the Dunford-Pettis property need their projective tensor product $X \hat{\otimes} Y$?

With regards to Question 9 we ought to mention the following characterization of spaces with the Dunford-Pettis property the proof of which is a direct consequence of Theorem 1 and the fact that the dual of the projective tensor product $X \hat{\otimes} Y$ is the space $\mathcal{L}(Y; X^*)$ of bounded linear operators from Y to X^*.

Theorem 16. *A Banach space X has the Dunford-Pettis property if and only if given a Banach space Y then $K \otimes J$ is relatively weakly compact in $X \hat{\otimes} Y$ whenever K and J are relatively weakly compact subsets of X and Y respectively.*

Actually the Dunford-Pettis property involves much better behavior of weakly compact sets than even Theorem 16 indicates. To see this we will describe a result of Ray Ryan [RR] which says that even polynomial operators on Dunford-Pettis spaces are nice. First, some notation.

For $k \geq 1$, $\mathcal{L}^k(X; Y)$ denotes the Banach space of all bounded k-linear maps $T: X^k \to Y$ with the norm

$$\|T\| = \sup\{\|T(x_1, \cdots, x_k)\|: \|x_1\|, \cdots, \|x_k\| \leq 1\}.$$

There is a canonical isomorphism of $\mathcal{L}^k(X; Y)$ with the space $\mathcal{L}(\hat{\otimes}^k X; Y)$ of bounded linear operators on the k-fold projective tensor product $\hat{\otimes}^k X$ to Y, given by associating with the k-linear map T the linear map $\hat{T}$, where

$$\hat{T}(x_1 \otimes \cdots \otimes x_k) = T(x_1, \cdots, x_k).$$

The k-linear map T is weakly compact if it maps bounded sets in X^k onto relatively weakly compact sets in Y. Since the closed unit ball of $\hat{\otimes}^k X$ is the closed absolutely convex hull of $B_X \otimes B_X \otimes \cdots \otimes B_X$, T is weakly compact if and only if the associated linear map is weakly compact.

The k-linear map T is completely continuous if given weakly Cauchy sequences $(x_n(1)), \cdots, (x_n(k))$ the sequence $(T(x_n(1), \cdots, x_n(k)))$ is norm convergent.

A Banach space X is said to have the polynomial Dunford-Pettis property if for every $k \geq 1$, weakly compact k-linear maps on X^k to any Y are completely continuous. It is clear from considering the case $k = 1$ that spaces with the polynomial Dunford-Pettis property have the Dunford-Pettis property. The elegant converse, also true, is the point of the next

Theorem 17 (R. Ryan). *If X has the Dunford-Pettis property then every weakly compact k-linear operator on X is completely continuous.*

Proof. Induction on k.

For $k = 1$ this is the definition of the Dunford-Pettis property; so assume we know the theorem for k-linear weakly compact operators and suppose T is a weakly compact $(k + 1)$ linear operator from X to Y. Let $(x_n(1)), \cdots, (x_n(k)), (x_n(k + 1))$ be weakly Cauchy sequences in X.

Case I. Suppose for some i: $1 \le i \le k + 1$ we have that $(x_n(i))$ is weakly null. Then for any $z_1, \cdots, z_k \in X^k$ we have

$$\|T(z_1, \cdots, z_{i-1}, x_n(i), z_i, \cdots, z_k)\| \to 0.$$

Hence T induces a k-linear map τ: $X^k \to c_0(Y)$ where the nth coordinate $\tau(z_1, \cdots, z_k)_n$ of $\tau(z_1, \cdots, z_k)$ is given by

$$\tau(z_1, \cdots, z_k)_n = T(z_1, \cdots, z_{i-1}, x_n(i), z_i, \cdots, z_k).$$

Of course τ is bounded and even weakly compact.

To see that τ is weakly compact we show the associated linear operator $\hat{\tau}$ from the k-fold projective tensor product $\hat{\otimes}^k X$ of X to $c_0(Y)$ is weakly compact. In order to do this we have to show that $\hat{\tau}^{**}\varphi$ belongs to $c_0(Y)$ for any $\varphi \in (\hat{\otimes}^k X)^{**}$; if things are looking complicated don't despair: we'll be out of the woods soon. Now $(\hat{\otimes}^k X)^*$ is nothing but $\mathcal{L}^k(X)$, the space of k linear functionals on X, so $(\otimes^k X)^{**}$ is $\mathcal{L}^k(X)^*$. We will avoid having to know the form of members of $\mathcal{L}^k(X)^* = (\otimes^k X)^{**}$. How? Well take a $\varphi \in \mathcal{L}^k(X)^*$ and define Φ: $X \to \mathcal{L}^{k+1}(X)^*$ by

$$\Phi(x)(S) = \varphi(S(\cdot, \ldots, \cdot, \underset{\text{ith slot}}{x}, \cdot, \ldots, \cdot)).$$

Observe that checking y^*'s in Y^* we have

$$\hat{T}^{**}(\Phi(x_n(i))) = \hat{\tau}^{**}(\varphi)_n.$$

Indeed, $y^*(\hat{\tau}^{**}(\varphi)_n) = \varphi((\hat{\tau}^* y^*)_n) = \varphi(y^* \hat{\tau}_n) = y^*\hat{T}(\Phi(x_n(i)))=\hat{T}^{**}\Phi(x_n(i))(y^*)$.

But $\hat{T}^{**}$ is weakly compact so that $\hat{T}^{**} \circ \Phi$ is a weakly compact linear operator from X into Y which must send the weakly null sequence $(x_n(i))$ to a norm null sequence since X has the Dunford-Pettis property. This though is just saying that $\hat{\tau}^{**}\varphi$ belongs to $c_0(Y)$ for each $\varphi \in \mathcal{L}^k(X)^* = (\hat{\otimes}^k X)^*$, that is, τ is weakly compact.

Our inductive hypothesis comes on the scene to warn us that $\tau(x_n(1),\dots, x_n(i-1), x_n(i+1),\dots, x_n(k+1))$ is a norm convergent sequence in $c_0(Y)$. It follows that the sequence $(T(x_n(1),\dots, x_n(i-1), x_n(i), x_n(i+1),\dots, x_n(k+1)))$, which is just the sequence of nth coordinates of the $c_0(Y)$-convergent sequence $\tau(x_n(1),\dots, x_n(i-1), x_n(i+1),\dots, x_n(k+1))$, must itself be norm convergent in Y, closing out Case I.

<u>Case II</u>. If we have that $(x_n(1)),\dots, (x_n(k+1))$ are all weakly Cauchy then for any strictly increasing sequences (k_n), (j_n) of positive integers each of the sequences $(x_{k_n}(i) - x_{j_n}(i))$, $1 \le i \le k+1$, is weakly null. Moreover,

$$T(x_{k_n}(1),\dots, x_{k_n}(k+1)) - T(x_{j_n}(1),\dots,x_{j_n}(k+1))$$

$$= \Sigma_{\ell=1}^{k+1} T(x_{k_n}(1),\dots,x_{k_n}(\ell-1), x_{k_n}(\ell) - x_{j_n}(\ell), x_{j_n}(\ell+1),\dots,x_{j_n}(k+1))$$

a finite sum each of whose terms is by Case I part of a norm null sequence. It follows that the sequence

$$(T(x_n(1),\dots,x_n(k+1)))$$

is norm convergent. This finishes the proof of the theorem.

Ryan does more in [RR] than just prove Theorem 17; he introduces the "holomorphic Dunford-Pettis property" and shows that the Dunford-Pettis property implies the holomorphic Dunford-Pettis property. If one wishes to appreciate the subtlety of Ryan's work one need only consider the efforts extended by A. Pelczynski [PP] in proving Theorem 17 for $X = C(K)$, K a dispersed compact Hausdorff space.

In addition to tensor product-like spaces there are other much more complicated vector-valued function spaces and in connection with these many Dunford-Pettis questions obviously remain open. For instance, when does $L_\infty(\mu, X)$ have the Dunford-Pettis property? When does the space $\mathcal{L}(X)$ of all bounded linear operators on X have the Dunford-Pettis property? More generally, when does $\mathcal{L}(X; Y)$ the space of all bounded linear operators from X to Y have the Dunford-Pettis property?

Herein there are several obvious problems. The first (and most obvious) is the inability to give any description of the dual space. What is the dual of $L_\infty(\mu, X)$? What is the dual of $\mathcal{L}(X)$? Even in case X is ℓ_2, no body knows what $\mathcal{L}(X)^*$ looks like!! A second difficulty encountered is the fact that the spaces involved are much bigger than the previously described vector-valued function spaces and as such can find themselves with unusually situated complemented subspaces.

To be more precise, what hypotheses might a real optimist hope would suffice to conclude that $L_\infty(\mu, X)$ have the Dunford-Pettis property? Clearly it'd be just that X have the Dunford-Pettis property. Is this enough? What about $\mathcal{L}(X; Y)$? Well here one would have to assume X^* and Y have the Dunford-Pettis property in order to conclude that $\mathcal{L}(X; Y)$ has it since both X^* and Y are complemented subspaces of $\mathcal{L}(X; Y)$. Is this enough? In each case, the answer is no.

Recall Stegall's example $X = (\Sigma \oplus \ell_2^n)_{\ell_1}$. This space enjoys the Schur property, yet its dual $(\Sigma \oplus \ell_2^n)_{\ell_\infty}$ fails the Dunford-Pettis property. Suppose we look at $L_\infty([0, 1], X)$. Plainly, $L_\infty([0, 1], X)$ contains a complemented subspace isomorphic to the ℓ_∞-sum of countably many copies of X. Look at a typical member of this ℓ_∞-sum. It consists of a sequence (x_n) of members of X and so corresponds to an array

$$x_1 = (\lambda_{11}^{(1)}, \lambda_{21}^{(1)}, \lambda_{22}^{(1)}, \lambda_{31}^{(1)}, \lambda_{32}^{(1)}, \lambda_{33}^{(1)}, \cdots)$$

$$x_2 = (\lambda_{11}^{(2)}, \lambda_{21}^{(2)}, \lambda_{22}^{(2)}, \lambda_{31}^{(2)}, \lambda_{32}^{(2)}, \lambda_{33}^{(2)}, \cdots)$$

$$x_3 = (\lambda_{11}^{(3)}, \lambda_{21}^{(3)}, \lambda_{22}^{(3)}, \lambda_{31}^{(3)}, \lambda_{32}^{(3)}, \lambda_{33}^{(3)}, \cdots)$$

$$\vdots$$

For any such (x_n) we can associate another sequence that is a member of the ℓ_∞ sum of countably many copies of X, the "diagonal" (dx_n) of (x_n):

$$dx_1 = (\lambda_{11}^{(1)}, 0, 0, 0, 0, 0, 0, 0, 0, 0, \cdots),$$

$$dx_2 = (0, \lambda_{21}^{(2)}, \lambda_{22}^{(2)}, 0, 0, 0, 0, 0, 0, 0, \cdots),$$

$$dx_3 = (0, 0, 0, \lambda_{31}^{(3)}, \lambda_{32}^{(3)}, \lambda_{33}^{(3)}, 0, 0, 0, 0, \cdots), \cdots .$$

The closed linear span of the sequences (dx_n) is easily seen to be complemented in the ℓ_∞-sum of countably many copies of X and so also in $L_\infty([0, 1], X)$. On the other hand a moment of thought shows this same closed linear span to be isomorphic to $(\Sigma \oplus \ell_2^n)_{\ell_\infty}$! So $L_\infty([0, 1], (\Sigma \oplus \ell_2^n)_{\ell_1})$ fails the Dunford-Pettis property even though $(\Sigma \oplus \ell_2^n)_{\ell_1}$ enjoys the Schur property. This fact was first shown me by Fred Delbaen and is due (I believe) to him.

Now $(\Sigma \oplus \ell_2^n)_{\ell_1}$ is the dual of $(\Sigma \oplus \ell_2^n)_{c_0}$ and so since $(\Sigma \oplus \ell_2^n)_{\ell_1}$ is a separable dual we know that

$$L_\infty([0, 1], (\Sigma \oplus \ell_2^n)_{\ell_1}) = \mathcal{L}(L_1[0, 1]; (\Sigma \oplus \ell_2^n)_{\ell_1}).$$

It follows that even though $L_1[0, 1]^*$ and $(\Sigma \oplus \ell_2^n)_{\ell_1}$ have the Dunford-Pettis property the space of operators $\mathcal{L}(L_1[0, 1]; (\Sigma \oplus \ell_2^n)_{\ell_1})$ fails the Dunford-Pettis property.

REFERENCES

[A] T. Ando, Convergent sequences of finitely additive measures, Pac. J. Math., 11 (1961), 395-404.

[KA] K. Andrews, Dunford-Pettis sets in the space of Bochner integrable functions, Math. Ann., 241 (1979), 35-42.

[JWB] J. W. Baker, Dispersed images of topological spaces and uncomplemented subspaces of C(X), Proc. Amer. Math. Soc. 41 (1973), 309-314.

[BDS] R. G. Bartle, N. Dunford and J. T. Schwartz, Weak compactness and vector measures, Canad. J. Math., 7 (1955), 289-305.

[B] J. Batt, Die Verallgemeinerungen des Darstellungssatzes von F. Riesz und ihre Anwendungen, Jber. Deutsch. Math. Verein, 74 (1973), 147-181.

[BP1] C. Bessaga and A. Pelczynski, On bases and unconditional convergence of series in Banach spaces, Studia Math 17 (1958), 151-168.

[BP2] C. Bessaga and A. Pelczynski, Spaces of continuous functions IV (On isomorphical classification of spaces C(S)), Studia Math. 19 (1960), 53-62.

[JB1] J. Bourgain, The Szlenk index and operators on C(K)-spaces, preprint.

[JB2] J. Bourgain, Dunford-Pettis operators on L^1 and the Radon-Nikodym property, preprint.

[JB3] J. Bourgain, An averaging result for ℓ^1-sequences and applications to weakly conditionally compact sets in L^1_X, preprint.

[JBDP] J. Bourgain, On the Dunford-Pettis property, preprint.

[BDW] J. Bourgain, J. Diestel and D. Weintraub, Limited sets in Banach spaces, preprint.

[JB] J. Brace, Transformations on Banach spaces, 1953 Cornell Ph.D. Dissertation.

[BD] J. K. Brooks and N. Dinculeanu, Weak compactness in spaces of Bochner integrable functions and applications, Advances in Math 24 (1977), 172-188.

[JC] J. Chaumat, Une généralisation d'un theoreme de Dunford-Pettis, Université de Paris XI, Orsay 1974.

[CD] I. Cnop and F. Delbaen, A Dunford-Pettis theorem for $L^1/H^{\infty\perp}$, J. Functional Analysis, 24 (1977), 364-378.

[JCL] J. Creekmore, Some contributions to the geometric study of the Lorentz function spaces, in preparation.

[D] F. Dashiell, Non-weakly compact operators on semicomplete C(S) lattices, with applications to Baire classes, preprint.

[DHH] F. Dashiell, A. Hager and M. Henriksen, Order-Cauchy completions of rings and vector lattices of continuous functions, preprint.

[DFJP] T.H.E. Davis, T. Figiel, W. B. Johnson and A. Pelczynski, Factoring weakly compact operators, J. Functional Analysis 17 (1974), 311-327.

[FD1] F. Delbaen, The Dunford-Pettis property for certain uniform algebras, Pacific J. Math. 65 (1976), 29-33.

[FD2] F. Delbaen, Weakly compact operators on the disk algebra, J. of ALGEBRA!! 45 (1977), 284-294.

[SIMP] J. Diestel, Remarks on weak compactness in $L_1(\mu, X)$, Glasgow Math J., 18 (1977), 87-91.

[DU] J. Diestel and J. J. Uhl, Jr., Vector Measures, American Mathematical Society's Mathematical Surveys Volume 15, 1977 Providence, R.I.

[DS] J. Diestel and C. J. Seifert, The Banach-Saks Operator Ideal, I: Operators on Spaces of Continuous Functions, Special Commemorative Issue of Commentationes Mathematical dedicated to Professor W. Orlicz's 75th Anniversary 1978-79.

[DOB] I. Dobrakov, On representation of linear operators on $C_0(T, X)$, Czech. Math. J., 21 (1971), 13-30.

[Dℓ_1] L. Dor, On sequences spanning a complex ℓ_1 space, Proc. Amer. Math. Soc., 47 (1975), 515-516.

[DPR] E. Dubinsky, A. Pelczynski and H. P. Rosenthal, On Banach spaces X for which $\pi_2(\mathcal{L}_\infty, X) = B(\mathcal{L}_\infty, X)$, Studia Math. 44 (1972), 617-648.

[DP] N. Dunford and B. J. Pettis, Linear operations on summable functions, Trans. Amer. Math. Soc. 47 (1940), 323-392.

[E$\ell\theta$] J. Elton, Ph.D. dissertation, Yale University 1978-79.

[F] H. Fakhoury, Sur les espaces de Banach ne contenant pas $\ell^1(N)$, Math. Scand. 41 (1977), 277-289.

[FBS] N. Farnum, The Banach-Saks theorem in C(S), Canad. J. Math. 26 (1974), 91-97.

[GLust] Y. Gordon and D. R. Lewis, Absolutely summing operators and local unconditional structures, Acta Math. 133 (1974), 27-48.

[GDP] A. Grothendieck, Sur les applications lineaires faiblement compartes d'espaces du type C(K), Canad. J. Math. 5 (1953), 129-173.

[GMEN] A. Grothendieck, Produits tensoriels Topologiques et Espaces Nucleaires, Memoirs Amer. Math. Soc. No. 16 (1955).

[GRes] A. Grothendieck, Résumé de la théorie métrique des produits tensoriels topologiques, Bol. Soc. Mat. Sao Paulo 8 (1956), 1-79.

[GTVS] A. Grothendieck, Espaces Vectorielles Topologiques, Inst. Mat. Pura Appl., Univ. de Sao Paulo, Sao Paulo, 1957.

[HAV] V. P. Havin, The spaces H^∞ and L^1/H^1_0, in: Studies in Linear Operators and Theory of Functions Volume 4, Nauchn. Sem. Leningrad. Otd. Mat. Inst. 29 (1974), 120-148.

[WBJ] W. B. Johnson, A complementably universal conjugate Banach space and its relation to the approximation problem, Israel J. Math. 13 (1972), 301-310.

[MON] W. B. Johnson, B. Maurey, G. Schechtman and L. Tzafriri, Symmetric Structures in Banach Spaces, Memoirs Amer. Math. Soc. No. 217 (1979).

[KP] M. I. Kadec and A. Pelczynski, Bases, lacunary sequences and complemented subspaces in the spaces L_p, Studia Math. 21 (1962), 161-176.

[KIS1] S. V. Kisliakov, On spaces with a "small" annihilator, in: Studies in Linear Operators and Theory of Functions, Volume 7, Nauchn. Sem. Leningrad Otd. Mat. Inst. 65 (1976), 192-195.

[KIS2] S. V. Kisliakov, Sobolev imbedding operators and the non-isomorphisms of certain Banach spaces, Functional Analysis and Applications, 9 (1976), 290-294.

[KIS3] S. V. Kisliakov, On the conditions of Dunford-Pettis, Pelczynski and Grothendieck, Dokl. Akad. Nauk SSSR 225 (1975), 1252-1255.

[KWP] S. Kwapien and A. Pelczynski, Some linear topological properties of the Hardy spaces H^p, Compositio Math. 33 (1976), 261-288.

[LLL_1] A. Lazar and J. Lindenstrauss, Banach spaces whose duals are L^1-spaces and their representing matrices, Acta Math. 126 (1971), 165-193.

[DRL1] D. R. Lewis, Integration with respect to vector measures, Pacific J. Math. 33 (1970), 157-165.

[DRL2] D. R. Lewis, Conditional weak compactness in certain inductive tensor products, Math. Ann. 201 (1973), 201-209.

[LP] J. Lindenstrauss and A. Pelczynski, Absolutely summing operators in $\mathcal{L}_p$ spaces and their applications, Studia Math. 29 (1968), 275-326.

[Loh] R. H. Lohman, A note on Banach spaces containing ℓ_1, Canad. Math. Bull. 19 (1976), 365-367.

[Lo] G. G. Lorentz, On the theory of the spaces Λ, Pacific J. Math. 1 (1951), 411-429.

[MS] S. Mazurkiewicz and W. Sierpinski, Contribution à la topologie des ensembles denombrables, Fund. Math. 1 (1920), 17-27.

[OR] E. Odell, Applications of Ramsey theorems to Banach space theory, preprint.

[PP^2] A. Pelczynski, Projections in certain Banach spaces, Studia Math, 19 (1960), 209-228.

[PV] A. Pelczynski, Banach spaces on which every unconditionally converging operator is weakly compact, Bull. Acad. Polon. Sci. 10 (1962), 641-648.

[PP] A. Pelczynski, On weakly compact polynomial operators on B-spaces with the Dunford-Pettis property, Bull. Acad. Polon. Sci. 11 (1963), 371-378.

[PS^2] A Pelczynski, On strictly singular and strictly co-singular operators:
I. Strictly singular and strictly co-singular operators in C(S)-spaces, Bull. Acad. Polon. Sci. 13 (1965), 31-36;
II. Strictly singular and strictly co-singular operators in $L(\nu)$-spaces, Bull. Acad. Polon. Sci. 13 (1965), 37-41.

[PKSU] A. Pelczynski, Banach Spaces of Analytic Functions and Absolutely Summing Operators, CBMS Regional Conference Series in Mathematics AMS publication number 30 (1976).

[PS] A Pelczynski and Z. Semadeni, Spaces of continuous functions. III. Spaces $C(\Omega)$ for Ω without perfect subsets, Studia Math. 18 (1959), 211-222.

[PSz] A. Pelczynski and W. Szlenk, An example of a non-shrinking basis, Revue Roum. Math. Pures et Appli. 10 (1965), 961-966.

[PT] P. Pethe and N. Thakare, A note on the Dunford-Pettis property and the Schur property, Indiana J. Math. 27 (1978).

[Ph] R. S. Phillips, On linear transformations, Trans. Amer. Math. Soc. 48 (1940), 516-541.

[HPRL] H. P. Rosenthal, On relatively disjoint families of measures, with some applications to Banach space theory, Studia Math. 37 (1970), 13-36.

[Rinj] H. P. Rosenthal, On injective Banach spaces and the spaces $L^\infty(\mu)$ for finite measures μ, Acta Math. 124 (1970), 205-248.

[Rℓ_1] H. P. Rosenthal, A characterization of Banach spaces containing ℓ_1, Proc. Nat. Acad. Sci. (USA) 71 (1974), 2411-2413.

[RR] R. Ryan, preprint (to appear in Bull. Acad. Polon. Sci.).

[PSθ] P. Saab, Integral representation by boundary vector measures, and applications, University of Illinois at Urbana-Champaign dissertation (1979).

[DS] D. Sarason, Functions of vanishing mean oscillation, Trans. Amer. Math. Soc. 207 (1975), 391-405.

[JS] J. Schreier, Ein Gegenbeispiel zur Theorie der schwachen Konvergenz, Studia Math. 2 (1930), 58-62.

[GS] G. L. Seever, Measures on F-spaces, Trans. Amer. Math. Soc. 133 (1968), 267-280.

[S] C. Stegall, Duals of certain spaces with the Dunford-Pettis property, Notices Amer. Math. Soc. 19 (1972), 799.

[CSℓ_1] C. Stegall, Banach spaces whose duals contain $\ell_1(\Gamma)$ with applications to the study of dual $L_1(\mu)$ spaces, Trans. Amer. Math. Soc. 176 (1973), 463-477.

[GT] G. E. F. Thomas, L'integration par rapport a une mesure de Radon vectorielle, Ann. Inst. Fourier (Grenoble) 20 (1970), 55-191.

Kent State University
Kent, Ohio 44242 (USA)

Contemporary Mathematics
Volume 2
1980

AN EXPANSION THEOREM

Dedicated to the Memory of

B. J. Pettis

by

Nelson Dunford

1 Introduction

The purpose of this note is to prove an expansion theorem for spectral operators T of finite type. It is shown that if such operators have $E(\sigma_c(T)) = 0$, where E is the spectral measure for T and $\sigma_c(T)$ is the continuous spectrum of T, then every vector x has a strongly convergent expansion $x = \sum E(\alpha_n)x$ with $\{\alpha_n\} \subseteqq \sigma_p(T)$, the point spectrum of T. If the spectrum of T is a countable set, it is not necessary to assume that $E(\sigma_c(T)) = 0$.

Section 2 explains the basic notation and terminology and Section 3 gives the necessary definitions. It also proves the ten results which are required before the expansion theorem is proved. Reference is made to Linear Operators by N. Dunford and J. T. Schwartz whenever possible. Theorems 13 and 17 in Section 3 are in Linear Operators. The proof given here of Theorem 13 is shorter and that of Theorem 17 is more lucid than in Linear Operators.

Some of the results concern the integral and are similar to those given in Linear Operators. This partial repetition is necessary, since the integral is defined differently than it was in Linear Operators.

Not only is the definition slightly different but the derivations of the properties of the integral in Linear Operators are different, in that use is made of XVIII.2.6 (such references are to Linear Operators). Also the definition of a spectral operator (XVIII.2.1) is more restrictive than the one given in this note. However, when reference is made to a proof in Linear Operators, the proof applies to the concepts as defined here.

2 Basic terminology

Throughout this note X is a complex Banach space and T is a lin-

Copyright © 1980, American Mathematical Society

ear operator whose domain $D(T)$ and range $R(T)$ are contained in X.

The operator T is said to be closed if $x_n \in D(T)$; $n = 1,2,\ldots$, $x_n \to x$ and $Tx_n \to y$ imply that x is in $D(T)$ and $Tx = y$. If T is closed and everywhere defined, it is in $B(X)$, the algebra of all bounded linear operators on X. This is proved in II.2.4.

Let $T(\alpha) \equiv \alpha I - T$ where I is the identity operator in X. Then, the point spectrum $\sigma_p(T)$ is the set of all α in the complex number system E^1 for which $T(\alpha)$ is not one-to-one on its domain, the residual spectrum $\sigma_r(T)$ consists of all α for which $T(\alpha)$ is one-to-one on its domain and the range $R(T(\alpha))$ is not dense in X, the continuous spectrum $\sigma_c(T)$ consists of all α for which $T(\alpha)$ is one-to-one on its domain, the range $R(T(\alpha))$ is dense in X and the inverse $T(\alpha)^{-1}$ is a discontinuous mapping of $R(T(\alpha))$ into $D(T(\alpha))$, the resolvent set $\rho(T)$ consists of all α for which $T(\alpha)$ is one-to-one on its domain, the range $R(T(\alpha))$ is dense in X and the inverse $T(\alpha)^{-1}$ is a continuous mapping of $R(T(\alpha))$ into $D(T(\alpha))$.

It is clear that $\sigma_p(T)$, $\sigma_r(T)$, $\sigma_c(T)$ and $\rho(T)$ are disjoint sets whose union is E^1. The spectrum $\sigma(T)$ of T is defined as $\sigma(T) \equiv \sigma_p(T) \cup \sigma_r(T) \cup \sigma_c(T) = \rho(T)'$ if this set is bounded; otherwise $\sigma(T) \equiv \rho(T)' \cup (\infty)$.

Since X is sequentially complete, there is, for every $\alpha \in \rho(T)$, a uniquely defined continuous linear operator $R(\alpha;T)$, called the resolvent of T at α, with domain X which is one-to-one and for which $R(\alpha;T)R(T(\alpha)) = D(T(\alpha))$, $(\alpha I - T)R(\alpha;T)x = x$; $x \in R(T(\alpha))$ and $R(\alpha;T)(\alpha I - T)x = x$; $x \in D(T(\alpha))$.

An equation using $\equiv$, such as

$$R(\alpha;T)x \equiv T(\alpha)^{-1}x; \quad x \in R(T(\alpha)),$$

means that the first term is defined by the second. This is the way one would start in proving the existence of the operator $R(\alpha;T) \in B(X)$. The symbol χ_σ is used for the characteristic function of σ, i.e., the function

$$\chi_\sigma(\alpha) = 1; \quad \alpha \in \sigma, \quad \chi_\sigma(\alpha) = 0; \quad \alpha \notin \sigma.$$

The symbol ϕ is used for the void set. If $X_0 \subseteqq D(T)$, the operator $T_0 \equiv T|X_0$ is the operator with domain $D(T_0) = X_0$ and for which $T_0x_0 = Tx_0$; $x_0 \in X_0$. The symbol $\{q \mid q \in Q, P(q)\}$ or $\{q \mid P(q)\}$ if Q is understood, is the set of all q in Q which satisfy the proposition P.

If $E(\sigma)$; $\sigma \in B$, the ring of all Borel sets in E^1, is a projection with $E(\sigma) \in B(X)$ which is additive on B, if $\sigma_1,\ldots,\sigma_n$ are disjoint sets in B and if $\alpha_1,\ldots,\alpha_n$ are complex numbers, it follows from IV.10.4(b) that

$$\text{(1)} \qquad \sum_{m=1}^{n} |\alpha_m E(\sigma_m)| \leqq 4 \sup_m |\alpha_m| \sup_\sigma |E(\sigma)| .$$

The symbol $|E|$ is defined by

$$\text{(2)} \qquad |E| \equiv \sup_\sigma |E(\sigma)| .$$

3 The principal theorems

1 Definition. A spectral mapping is a mapping $\sigma \longrightarrow E(\sigma)$ of B into B(X) which has the properties that for every μ,σ in B

(i) $E(\emptyset) = 0,$

(ii) $E(\mu \cap \sigma) = E(\mu)E(\sigma),$

(iii) $E(\mu \cup \sigma) = E(\mu) + E(\sigma) - E(\mu)E(\sigma),$

(iv) $|E| < \infty .$

Note. It is clear from (ii) that $E(\sigma)^2 = E(\sigma)$ and that $E(\sigma),E(\mu)$ commute, from (i), (ii) and (iii) that $E(\sigma)$ is additive in σ and from (iv) that $E(\sigma)x$ is continuous in x uniformly with respect to $\sigma \in B$.

2 Definition. A spectral measure is a spectral mapping with the property that for each x in X, $E(\sigma)x$ in weakly countably additive in σ .

3 Theorem (Pettis [2] (1,2). A weakly countably additive vector-valued set function v defined on a σ-field F is countably additive. In addition, if m is a finite positive measure on F and if v vanishes on m-null sets, then v is m-continuous.

Note. It follows from this theorem of Pettis that every spectral measure is strongly countably additive.

4 Definition. A linear operator T whose domain and range are contained in X is called a spectral operator if it is closed and if there is a spectral measure E such that for δ, σ, μ in B with δ bounded,

(i) $E(\delta)X \subseteqq D(T), \quad TE(\delta) \in B(X),$

(ii) $E(\sigma)D(T) \subseteqq D(T),$

(iii) $TE(\sigma)x = E(\sigma)Tx; \quad x \in D(T)$,

(iv) the spectrum $\sigma(T_\mu)$ of the restriction $T_\mu \equiv T|(D(T) \cap E(\mu)X)$ satisfies the inclusion

$$E^1 \cap \sigma(T_\mu) \subseteqq \bar{\mu} .$$

Note. It is shown in Theorem XVIII.2.5 that every spectral operator T determines uniquely a spectral measure E which establishes 4(i),..., 4(iv).

In the remaining part of this note, the symbols f, g with or without subscripts will always refer to Borel measurable mappings of E^1 into E^1 and $|f|_\infty \equiv \sup_\alpha |f(\alpha)|$.

5 Definition. Let f be bounded and f_n; $n = 1, 2, \cdots$ finitely valued with $|f - f_n|_\infty \rightarrow 0$. It follows from (1), (2) of Section 2 that

$$\text{(i)} \qquad \left|\int_\sigma f_m dE - \int_\sigma f_n dE\right| \leqq 4\,|f_m - f_n|_\infty |E| \rightarrow 0 ,$$

uniformly with respect to σ in B. The integrals are defined by the equations

$$\text{(ii)} \qquad \int_\sigma f dE = \lim_n \int_\sigma f_n dE; \quad \sigma \in B .$$

We use the symbol $f(S)$ for $\int_{E^1} f dE$ and, if $f(\alpha) = \alpha$; $\alpha \in E^1$, we write S instead of $f(S)$. These operators are called bounded scalar operators. It follows from the inequality (i) that the integrals (ii) are independent of the sequence used to define them. This inequality also shows that

$$\text{(iii)} \qquad \left|\int_\sigma f dE\right| \leqq 4|f|_\infty |E| .$$

6 Definition. For every f and every spectral measure E, the operator $f(S)$ is defined as follows. Let

$$\text{(i)} \qquad \mu(n) \equiv \{\alpha \mid |f(\alpha)| \leqq n\}, \quad f_n \equiv f\chi_{\mu(n)} ,$$

$$\text{(ii)} \quad D(f(S)) \equiv \{x \mid \lim_n f_n(S)x \text{ exists and } \lim_n E(\mu(n))x = x\} .$$

Then

$$\text{(iii)} \qquad f(S)x \equiv \lim_n f_n(S)x; \quad x \in D(f(S)) .$$

The integrals are defined by

$$\text{(iv)} \quad \int_\sigma f dEx \equiv E(\sigma)f(S)x = \lim_n \int_\sigma f_n dEx; \; x \in D(f(S)), \; \sigma \in B.$$

The operators $f(S)$ are called scalar operators and, as before, we write S for $f(S)$ in case $f(\alpha) = \alpha$, $\alpha \in E^1$.

Note. It will sometimes be convenient to write $\int f(\alpha)E(d\alpha)x$ instead of $\int f dEx$. Also, if f,g are bounded, then $f(S)g(S) = (fg)(S)$. This is clear if f,g are characteristic functions. Definition 5 shows that it holds for all bounded f and g. Thus, if $(\beta - \alpha)^{-1}$ is bounded in α except on a set for which E vanishes, we have

$$(\beta I - S)\int(\beta - \alpha)^{-1}E(d\alpha) = I.$$

7 Theorem. Every scalar operator is closed.

Proof. Let $f(S)$ be the operator of Definition 6 and let $\mu(n)$ be the set defined therein. To see that $f(S)$ is closed, let $x_m \in D(f(S))$, $\lim_m x_m = x$, $\lim_m f(S)x_m = y$. Then

$$y = \lim_m f(S)x_m = \lim_m \lim_n E(\mu(n))f(S)x_m.$$

Since

$$|E(\mu(n))y - E(\mu(n))f(S)x_m| \leqq |E||y - f(S)x_m|,$$

it follows that

$$\lim_m E(\mu(n))f(S)x_m = E(\mu(n))y, \text{ uniformly in } n.$$

It follows from a theorem of E. H. Moore (I.7.6) that the following limits exist and are equal to y.

$$\lim_{n,m} E(\mu(n))f(S)x_m, \quad \lim_n E(\mu(n))y.$$

Also, since $\lim_m E(\mu(n))x_m = E(\mu(n))x$ uniformly in n, the theorem of E. H. Moore shows that $\lim_n E(\mu(n))x = x$, which completes the proof. ∎

8 Theorem. The scalar operator $f(S)$ is a spectral operator whose spectral measure is

(i) $$E(\mu; f(S)) = E(f^{-1}(\mu)); \quad \mu \in B.$$

Proof. To establish the condition 4(i), let σ be a bounded Borel set and $E(f^{-1}(\sigma))x = x$. Then, if $\mu(n)$ is the set of Definition 6, it follows that

$$f_n(S)x = \int_{f^{-1}(\sigma)} f_n dEx = \int_{f^{-1}(\sigma)\cap\mu(n)} f dEx = \int_{f^{-1}(\sigma)} f dEx,$$

for sufficiently large n, which establishes 4(i).

To establish 4(ii), note that

$$f_n(S)E(f^{-1}(\sigma))x = E(f^{-1}(\sigma))f_n(S)x; \quad x \in D(f(S)), \ \sigma \in B.$$

This shows that $E(f^{-1}(\sigma))D(f(S)) \subseteqq D(f(S))$ and that

$$f(S)E(f^{-1}(\sigma))x = E(f^{-1}(\sigma))f(S)x; \quad x \in D(f(S)), \quad \sigma \in B,$$

which establishes 4(ii) and 4(iii).

It is clear that $E(f^{-1}(\mu))$ is countably additive on B. Theorem 7 shows that $f(S)$ is closed.

It remains only to establish 4(iv), i.e., the inclusion

$$(1) \qquad E^1 \cap \sigma(f(S)|(D(f(S)) \cap E(f^{-1}(\mu))X)) \subseteq \bar{\mu}; \quad \mu \in B .$$

Let $x = E(f^{-1}(\mu))x$ and let $\alpha \notin \bar{\mu}$ so that the operator

$$R \equiv \int_{f^{-1}(\mu)} (\alpha - f(\beta))^{-1} E(d\beta)$$

is in $B(X)$. Thus, if x is also in $D(f(S))$,

$$(2) \qquad R(\alpha I - f(S))x = \lim_m \int_{\mu(n)} (\alpha - f(\beta))^{-1}(\alpha - f(\beta))E(d\beta)E(f^{-1}(\mu))x$$

$$= \int_{E^1} E(d\beta)E(f^{-1}(\mu))x = E(f^{-1}(\mu))x = x .$$

Since α is an arbitrary point in $\bar{\mu}'$, it follows from (2) that

$$\bar{\mu}' \subseteq \rho(f(S)|(D(f(S)) \cap E(f^{-1}(\mu))X)) ,$$

which establishes (1) and completes the proof of (i). ∎

9 Theorem. The spectral measure E of a spectral operator T has

$$E(E^1) = I = E(\sigma(T) \cap E^1), \; E(\rho(T)) = 0 .$$

Proof. Let $\sigma(n) \equiv \{\alpha \mid |\alpha| \leq n\}$ and $T_{\sigma(n)} \equiv T|E(\sigma(n))X$. It follows from Lemma XVIII.2.2 that $T_{\sigma(n)}$ is a spectral operator whose spectral measure $E_{\sigma(n)}$ satisfies the identity

$$(1) \qquad E_{\sigma(n)}(\sigma)E(\sigma(n)) = E(\sigma \cap \sigma(n)); \quad \sigma \in B .$$

Corollary XV.3.5 shows that $E_{\sigma(n)}(\rho(T_{\sigma(n)})) = 0$ and, since $\rho(T) \subseteq \rho(T_{\sigma(n)})$, it follows that $E_{\sigma(n)}(\rho(T)) = 0$. It follows from (1) that $E(\rho(T) \cap \sigma(n)) = 0$ and the conclusions follow from the countable additivity of $E(\cdot)x$. ∎

<u>Note</u>. Theorem 8 shows that $f(S)$ is a spectral operator. Thus Theorem 9 shows that the requirement that $E(\mu(n))x \to x$ in the definition of the domain $D(f(S))$, as stated in Definition 6(ii), is redundant.

10 Corollary. **Every spectral operator is densely defined.**

Proof. Let T be a spectral operator with spectral measure E and let $\{\sigma(n)\}$ be a sequence of bounded Borel sets with $\sigma(n) \uparrow E^1$. It follows from Definition 4(i) that $E(\sigma(n))x \in D(T); \; x \in X, \; n = 1,2,\ldots$, and from Theorem 9 and the countable additivity of $E(\cdot)x$ that

$E(\sigma(n))x \to x; \quad x \in X$. ∎

The following theorem shows that the integral of Definition 6 is independent of the sequence $\{\mu(n)\}$ used to define it. This observation justifies the use of the term <u>integral</u>.

11 Theorem. Let E be a spectral measure, $x \in D(f(S))$ and

(i) $$\nu(\sigma) \equiv \int_\sigma f dEx = E(\sigma)f(S)x; \quad \sigma \in B .$$

Then ν is countably additive on B and independent of the sequence $\{\mu(n)\}$ used to define it, in the sense that if $\{\sigma(n)\} \subset B$, if $\sigma(n) \uparrow E^1$ and if the functions $g_n \equiv f\chi_{\sigma(n)}$; $n = 1,2,\cdots$ are bounded, then

(ii) $$D(f(S)) = \{x \mid \lim_n g_n(S)x \text{ exists}\} ,$$

and

(iii) $$\nu(\sigma) = E(\sigma) \lim_n g_n(S)x; \quad x \in D(f(S)), \quad \sigma \in B .$$

Proof. Let $\sigma_n \downarrow \emptyset$. The condition 1(iv) shows for $x \in D(f(S))$,

$$\nu(\sigma_n \sigma) = E(\sigma)E(\sigma_n)f(S)x \to 0$$

uniformly with respect to σ in B . Thus, since $\mu(n)' \downarrow \emptyset$ and $\sigma(n)' \downarrow \emptyset$ we see that

$$E(\sigma)g_n(S)x - E(\sigma)f_n(S)x = \nu(\sigma\,\sigma(n)\mu(n)') - \nu(\sigma\,\sigma(n)'\mu(n)) \to 0$$

uniformly with respect to σ in B and thus the limits $\lim_n E(\sigma)g_n(S)x$; $\sigma \in B$ exist if and only if the limits $\lim_n E(\sigma)f_n(S)x$; $\sigma \in B$ exist and when they exist, they are equal. ∎

12 Definition. An operator N is said to be <u>nilpotent</u> if $N \in B(X)$ and $N^n = 0$ for some positive integer n . If $N^m \neq 0$ and $N^{m+1} = 0$, then m is said to be the <u>order</u> of N . If $N = 0$, we define $N^0 \equiv I$ so that the <u>order of</u> 0 is 0 . If S is a scalar operator with spectral measure E and if N is a nilpotent operator with

(i) $$E(\sigma)N = NE(\sigma); \quad \sigma \in B ,$$

then the operator $T \equiv S + N$ is said to be an operator of <u>finite type</u> with <u>scalar part</u> S and <u>nilpotent part</u> N .

<u>Note</u>. In all following discussions which involve an operator T of finite type, the symbols S, E, N, T are related as they are in this definition, and N is always a nilpotent operator.

13 Theorem. Every operator of finite type is a spectral operator whose spectral measure is that of its scalar part.

Proof. Since $N \in B(X)$, we have $D(S) = D(T)$, and thus it is clear that T satisfies the conditions 4(i), (ii), (iii). To establish 4(iv), let $\alpha \notin \bar{\mu}$ and let $T_\mu \equiv T \mid (D(T) \cap E(\mu)X)$, $S_\mu \equiv S \mid (D(S) \cap E(\mu)X)$, $N_\mu \equiv N \mid E(\mu)X$, $I_\mu \equiv I \mid E(\mu)X$, so that 4(iv) applied to S shows that $\alpha \in \rho(S_\mu)$. Let $R_\mu \equiv R(\alpha; S_\mu)$, $N^{m+1} = 0$ and $P \equiv \sum_{n=0}^{m} N_\mu^n R_\mu^{n+1}$. Then

$$(\alpha I_\mu - T_\mu)P = (\alpha I_\mu - S_\mu - N_\mu)P = \sum_{n=0}^{m} N_\mu^n R_\mu^n - \sum_{n=0}^{m} N_\mu^{n+1} R_\mu^{n+1} = I_\mu .$$

Thus $\alpha \in \rho(T_\mu)$ and, since α is an arbitrary point in $\bar{\mu}'$, it follows that $\bar{\mu}' \subseteq \rho(T_\mu)$, which establishes 4(iv) for T. ∎

14 Corollary. If E is the spectral measure for the operator T of finite type and if $\{\sigma_n\}$ is any sequence of bounded Borel sets with $\sigma_n \uparrow E^1$, then

$$D(T) = \{x \mid \lim_n TE(\sigma_n)x \text{ exists}\}.$$

Proof. Since $N \in B(X)$, we have $D(S) = D(T)$ and the limit $\lim_n TE(\sigma_n)x$ exists if and only if the limit $\lim_n SE(\sigma_n)x$ exists. The conclusion follows from Theorem 11. ∎

15 Theorem. Let T be a spectral operator of finite type and let E be its spectral measure. Then

(i) $\beta \in \sigma_p(T)$ if and only if $E(\beta) \neq 0$,

(ii) if $\beta \in \sigma(T)$, then $\beta \in \sigma_c(T)$ if and only if $E(\beta) = 0$,

(iii) $\sigma_r(T) = \emptyset$.

Proof. To prove (i) let $E(\beta)x = x \neq 0$. Since $SE(\beta) = \int_{(\beta)} \alpha E(d\alpha) = \beta E(\beta)$, we have $-NE(\beta) = (\beta I - T)E(\beta)$ and $(-1)^n N^n E(\beta) = (\beta I - T)^n E(\beta)$. Thus

$$(\beta I - T)^n x = (-1)^n N^n x; \quad n = 0, 1, 2, \cdots .$$

This shows that there is an integer p in the range $0 \leqq p \leqq m$, where m is the order of N such that $(\beta I - T)^{p+1}x = 0$ and $(\beta I - T)^p x \neq 0$. If $y \equiv (\beta I - T)^p x$, we have

$$(1) \qquad (\beta I - T)y = 0 \text{ and } y \neq 0 ,$$

so that $\beta \in \sigma_p(T)$. Now, conversely, if $y \in D(T)$ and satisfies (1), let $\sigma(n) = \{\alpha \mid |\beta - \alpha| \geqq n^{-1}\}$ and $T_{\sigma(n)} \equiv T \mid E(\sigma(n))X$. It follows from 4(iv) that $\beta \in \rho(T_{\sigma(n)})$ and thus

$$E(\sigma(n))y = R(\beta; T_{\sigma(n)})(\beta I - T)E(\sigma(n))y = R(\beta; T_{\sigma(n)})E(\sigma(n))(\beta I - T)y = 0 ,$$

and $(I - E(\beta))y = \lim_n E(\sigma(n))y = 0$, $E(\beta)y = y \neq 0$, which proves (i).

To prove (ii) we use a theorem of Foguel [1] which states that if

$T \in B(X)$ and if $E(\beta) = 0$, then $(\beta I - T)X$ is dense in X. Let $\sigma(n) \equiv \{\alpha \mid |\alpha| \leq n\}$ and let $E(\beta) = 0$ so that 4(i) and the result of Foguel show that $(\beta I - TE(\sigma(n)))X$ is dense in X and thus $(\beta I - TE(\sigma(n)))E(\sigma(n))X$ is dense in $E(\sigma(n))X$. Let

$$X_n \equiv (\beta I - TE(\sigma(n)))E(\sigma(n))X = (\beta I - T)E(\sigma(n))X$$

so that X_n is dense in $E(\sigma(n))X$. Thus $\bigcup_{n=1}^{\infty} X_n$ is dense in $\bigcup_{n=1}^{\infty} E(\sigma(n))X$ which, by Theorem 9 and the countable additivity of $E(\cdot)x$ is dense in X. By 4(i) we have

$$\bigcup_{n=1}^{\infty} X_n = (\beta I - T) \bigcup_{n=1}^{\infty} E(\sigma(n))X \subseteq (\beta I - T)D(T),$$

which shows that $(\beta I - T)D(T)$ is dense in X. Thus, $\beta \notin \sigma_r(T)$ and, since $E(\beta) = 0$, the conclusion (i) shows that $\beta \notin \sigma_p(T)$. Thus, if $E(\beta) = 0$ and $\beta \in \sigma(T)$, we have $\beta \in \sigma_c(T)$. The conclusion (iii) follows from (i) and (ii). ∎

16 Theorem. If T is a spectral operator of finite type with scalar part S and spectral measure E, then

(i) $\sigma_p(T) = \sigma_p(S)$ and this set consists of all α with $E(\alpha) \neq 0$,

(ii) $\sigma_c(T) = \sigma_c(S)$ and consists of all $\alpha \in \sigma(S)$ with $E(\alpha) = 0$,

(iii) $\sigma_r(T) = \sigma_r(S) = \emptyset$,

(iv) $\sigma(T) = \sigma(S)$, $\rho(T) = \rho(S)$,

(v) $\sigma(T)$ is closed, $\rho(T)$ is open.

Proof. Since S is a spectral operator of finite type with nilpotent part 0, the conclusions (i) and (iii) follow from Theorems 13 and 15. The conclusion (iv) for bounded T follows from Theorem XV.4.5 and for unbounded T from Theorem 13 and Lemma XVIII.2.25. Lemma XVIII.2.25 also shows that $\sigma(T)$ is closed and thus $\rho(T)$ is open. The conclusion (ii) is an immediate consequence of the other conclusions. ∎

17 Theorem. Let T be a spectral operator with spectral resolution E, and let the space X be separable. Then the set $\{\alpha \mid E(\alpha) \neq 0\}$ is countable.

Proof. For each α with $E(\alpha) \neq 0$, let x_α be a vector satisfying the equations

$$(1) \qquad x_\alpha = E(\alpha)x_\alpha, \quad |x_\alpha| = 1.$$

Then, for two different numbers α, β with $E(\alpha) \neq 0$, $E(\beta) \neq 0$, we have $E(\alpha)E(\beta) = E((\alpha) \cap (\beta)) = E(\emptyset) = 0$ and

$$\begin{aligned} 1 = |x_\alpha| &= |E(\alpha)x_\alpha| \\ &= |E(\alpha)(E(\alpha)x_\alpha - E(\beta)x_\beta)| \\ &\leqq |E||x_\alpha - x_\beta| . \end{aligned}$$

Thus

$$(2) \qquad |x_\alpha - x_\beta| \geqq \frac{1}{|E|} ; \quad \alpha \neq \beta .$$

Let $\{x_n\}$ be dense in X. Since $|x|$ is continuous in x, the equation (2) shows that there is a positive number ε such that

$$(3) \quad |x_p - x_q| \geqq \frac{1}{2|E|} ; \ |x_p - x_\alpha| < \varepsilon , \ |x_q - x_\beta| < \varepsilon , \quad \alpha \neq \beta .$$

For each α with $E(\alpha) \neq 0$, let (α, ε) be the set of all x_m for which $|x_m - x_\alpha| < \varepsilon$. Then (3) gives

$$(4) \qquad (\alpha, \varepsilon) \cap (\beta, \varepsilon) = \emptyset ; \quad \alpha \neq \beta .$$

The equation (4) shows that the numbers α in the set $\{\alpha | E(\alpha) \neq 0\}$ are in a one-to-one correspondence with mutually disjoint subsequences of $\{x_n\}$, which proves that the set $\{\alpha | E(\alpha) \neq 0\}$ is countable. ∎

18 Theorem. Let T be a spectral operator of finite type and with spectral resolution E. It is also assumed that

$$(i) \qquad \sigma_c(T) \in B, \quad E(\sigma_c(T)) = 0.$$

Then

(ii) for each x in X there is a countable set $\{\alpha_n\} \subseteqq \sigma_p(T)$ such that

$$x = \sum_n E(\alpha_n)x .$$

(iii) If X is separable, the countable set $\{\alpha_n\}$ may be chosen independent of x. It consists of all α for which $E(\alpha) \neq 0$ and

$$x = \sum_n E(\alpha_n)x; \quad x \in X .$$

Proof. Consider first the case where X is separable. It follows from Theorems 9, 16 and the hypothesis (i) that $E(\sigma_p(T) = I$. Theorems 16(i) and 17 show that the point spectrum is a countable set $\{\alpha_n\}$. Thus the conclusion (iii) follows from the countable additivity of $E(\cdot)x$.

Let B_o be the family of bounded sets in B and $X(x)$ the closed linear manifold in X which is determined by the set $\{T^n E(\mu)x;\ \mu \in B_o, n = 0, 1, 2, \cdots\}$. Let

$$(1) \quad T_x \equiv T|(D(T) \cap X(x)), \quad E_x(\sigma) \equiv E(\sigma)|X(x), \quad \sigma \in B .$$

It will be shown that T_x is a spectral operator in $X(x)$ whose spectral resolution is E_x. Let $z_n \in D(T_x) = D(T) \cap X(x)$, $z_n \to z$, $T_x z_n \to y$. Then $T_x z_n = Tz_n \to y$, and since T is closed, z is in $D(T)$ and $Tz = y$. Since $X(x)$ is closed, z is in $X(x)$ so that $z \in D(T) \cap X(x) = D(T_x)$ and $T_x z = Tz = y$, which proves that T_x is closed.

Let $y \in X(x)$, $\{\sigma_n\} \subset B_o$, $\sigma_n \uparrow \sigma$. Then $E(\sigma_n)y \in X(x)$, $y = \lim E(\sigma_n)y \in X(x)$ which shows that

$$(2) \qquad E(\sigma)X(x) \subseteq X(x); \quad \sigma \in B .$$

Also Corollary 14 shows that if $y \in D(T) \cap X(x)$, if $\{\sigma_n\} \subset B_o$ and $\sigma_n \uparrow E^1$, we have $Ty = \lim TE(\sigma_n)y \in X(x)$. Thus

$$(3) \qquad T_x \text{ maps } D(T) \cap X(x) \text{ into } X(x) .$$

Let σ be a bounded set in B. Then $E_x(\sigma)X(x) = E(\sigma)X(x) \subseteq D(T) \cap X(x) = D(T_x)$, which establishes 4(i).

Let $z \in D(T) \cap X(x)$ so that 4(ii) shows that $E(\sigma)z \in D(T)$. This and (2) show that

$$E_x(\sigma)z = E(\sigma)z \in D(T) \cap X(x) = D(T_x); \quad \sigma \in B, \quad z \in D(T_x),$$

which establishes 4(ii). For z in $D(T_x)$ we also have

$$T_x E_x(\sigma)z = TE(\sigma)z = E(\sigma)Tz = E_x(\sigma)T_x z; \quad \sigma \in B ,$$

which establishes 4(iii).

To establish 4(iv) for T_x let $T_\mu \equiv T|(D(T) \cap E(\mu)X)$, $S_\mu \equiv S|(D(S) \cap E(\mu)X)$, $N_\mu \equiv N|E(\mu)X$, $I_\mu \equiv I|E(\mu)X$ and let $T_{x\mu} \equiv T_x|(D(T_x) \cap E_x(\mu)X(x))$. It will first be shown that

$$(4) \qquad \rho(T_\mu) \subseteq \rho(T_{x\mu}); \quad \mu \in B .$$

Let $\beta \in \rho(T_\mu)$. It follows from (2), (3), Corollary 14 and the definition of $X(x)$ that

$$(\beta I_\mu - T_\mu)(D(T_\mu) \cap E(\mu)X(x)) \subseteq E(\mu)X(x); \quad \mu \in B .$$

Thus, if

$$(4)' \qquad R(\beta; T_\mu)E(\mu)X(x) \subseteq E(\mu)X(x); \quad \mu \in B ,$$

it follows that the restriction of $R(\beta; T_\mu)$ to $E(\mu)X(x)$ is the resolvent $R(\beta; T_{x\mu})$ of the restriction $T_{x\mu}$, i.e., $\beta \in \rho(T_{x\mu})$. Thus, to establish (4), it suffices to prove (4)'.

To establish (4)' it is first assumed that $N = 0$ so that $T = S$. Theorem 9 shows that $E(\rho(S_\mu)) = 0$ and, since $\sigma(S_\mu)$ is closed, the function $(\beta - \alpha)^{-1}$ is bounded on $\sigma(S_\mu)$. Thus it follows from Definition 5 that

(5) $R(\beta;S_\mu)E(\mu)y = \int_{E^1}(\beta-\alpha)^{-1}E_\mu(d\alpha)E(\mu)y \in E(\mu)X(x);\ y \in X(x),$

which proves (4)' in case $T = S$. Now if $N \neq 0$ and $N^{m+1} = 0$, let $R_\mu \equiv R(\beta;S_\mu)$ and define

(6) $$P_\mu \equiv \sum_{n=0}^{m} N_\mu^n R_\mu^{n+1},$$

so that

(7) $$(\beta I_\mu - T_\mu)P_\mu = (\beta I_\mu - S_\mu - N_\mu)P_\mu$$
$$= \sum_{n=0}^{m} N_\mu^n R_\mu^n - \sum_{n=0}^{m} N_\mu^{n+1} R_\mu^{n+1} = I_\mu ,$$

which shows that

(8) $$P_\mu = R(\beta;T_\mu).$$

If $\sigma \in B_0$ and $y \in X(x)$, then $SE(\sigma)y = \int_\sigma \alpha E(d\alpha)y$. Thus (2) and the definition of the integral show that $SE(\sigma)y \in X(x);\ y \in X(x)$. It follows from the definition of $X(x)$ that $TE(\sigma)y \in X(x);\ y \in X(x)$. Thus $NE(\sigma)y \in X(x);\ y \in X(x),\ \sigma \in B_0$. Now let $\{\sigma_n\} \subset B_0$, $\sigma_n \uparrow E^1$. Theorem 9 shows that

(9) $$N_\mu E(\mu)y = \lim_n N_\mu E(\mu\sigma_n)y \in E(\mu)X(x);\quad y \in X(x).$$

Equations (5), (6), (8) and (9) show that $R(\beta;T_\mu)$ maps $E(\mu)X(x)$ into itself, which establishes (4)'. By taking complements in E^1, it is seen from (4) and Definition 4(iv) applied to T, that for every μ in B,

$$E^1 \cap \sigma(T_x \mid (D(T_x) \cap E_x(\mu)X(x))) \subseteq E^1 \cap \sigma(T \mid (D(T) \cap E(\mu)X)) \subseteq \bar{\mu},$$

which proves that T_x satisfies 4(iv).

This completes the proof that T_x is a spectral operator in $X(x)$ whose spectral resolution is E_x.

The following proof of the separability of $X(x)$ is due to William G. Bade. Let B_1 denote the countable subfield of B which is generated by the closed rectangles whose corners have rational co-ordinates, and let Y be the closed linear manifold determined by the set $\{E(\mu)x \mid \mu \in B_1\}$. It follows from Lemma XVII.3.12 that there is a linear functional x^* in X^* with the properties

(10) $x^*E(\sigma)x \geqq 0;\ \sigma \in B,$

(11) if $x^*E(\sigma)x = 0$ for some σ in B, then $E(\sigma)x = 0$,

(12) $x^*E(\cdot)x$ is a regular measure.

If μ is any open set, there is a non-decreasing sequence $\{\mu_n\}$ in B_1

with $\mu_n \uparrow \mu$ and so $E(\mu)x = \lim_n E(\mu_n)x$ is in Y. If μ is any Borel set, we use (12) to choose a non-increasing sequence $\{\mu_n\}$ of open sets with $\mu_n \supseteq \mu$; $n = 1,2,\ldots$ and such that $\lim_n x^*(\mu_n - \mu)x = 0$. Equations (10), (11) and Theorem 3 show that $E(\mu)x = \lim_n E(\mu_n)x$, which proves that $E(\mu)x \in Y$; $\mu \in B$, and this shows that $X(x)$ is separable.

Since $X(x)$ is separable, Theorems 16 and 17 show that T as an operator on $D(T) \cap X(x)$ has a countable set $\{\alpha_n\}$ for which $E(\alpha) \neq 0$ and

(13) $\{\alpha_n\} = \sigma_p(T_x) \subseteqq \sigma_p(T)$,

(14) $\sigma_r(T) = \emptyset$.

Next note that equations (13), (14), Theorem 9, hypothesis (i) and equation (1) show that

$$x = E_x(E^1)x = E_x(\{\alpha_n\})x = E(\{\alpha_n\})x ,$$

and the conclusion (ii) follows from the countable additivity of $E(\cdot)x$.

■

References

1 Foguel, S. R. The relations between a spectral operator and its scalar part. Pacific J. Math. 8, 51-65 (1958).

2 Pettis, B. J.

1. On integration in vector spaces. Trans. Amer. Math. Soc. 44, 277-304 (1938).
2. Absolutely continuous functions in vector spaces. (Abstract) Bull. Amer. Math. Soc. 45, 677 (1939).

Contemporary Mathematics
Volume 2
1980

THE RADON-NIKODYM PROPERTY

R. Huff

1. INTRODUCTION.

In August 1975 there was a special session on the Radon-Nikodym Property (RNP) at the American Mathematical Society meeting in Kalamazoo, Michigan. At that meeting, Professor B.J. Pettis gave a spellbinding talk on the early history of the subject. His story began with the interesting method by which J.D. Tamarkin found his way to the United States (see [7, p.208].) It was Tamarkin who suggested that J.A. Clarkson look at differentiability properties of vector-valued functions; this led to Clarkson's fundamental paper [5].

In that 1936 paper, Clarkson introduced the notion of a uniformly convex (UC-) space, and proved that a UC-space has the RNP. The paper is best known for the inequalities which he established in order to prove that every L^p $(1 < p < \infty)$ is UC.

It was soon shown by Pettis [25], and independently by D.P. Milman [24], that every UC-space is reflexive; and then as one of many results in a major paper, N. Dunford and B.J. Pettis [8] showed that reflexive spaces have the RNP. In fact, they proved the more fundamental result that every separable dual space has the RNP (see below). This sufficient condition for the RNP is the cornerstone of the entire area of research; it has been and remains the major tool in applications.

More than thirty years later, H. Maynard [22], proving a converse of a theorem of M. Rieffel [29], established a completely geometric characterization

Copyright © 1980, American Mathematical Society

of the RNP. This led to an intense study of the RNP in the 1970's and a great number of characterizations of the RNP were obtained. The results of this study, through 1976, are reported in the book of J. Diestel and J.J. Uhl [7].

Since the subject of the RNP is so well expounded in Diestel and Uhl's book, it makes no sense to attempt even a modest survey of the subject here. Instead, I shall attempt to simply whet the appetite of the reader who I assume is new to the subject.

In order to show intuitively the connection between the analytically defined RNP (see below) and the geometry of the space, we shall present in §2 a proof of the classical scalar case of the Radon-Nikodym theorem, and deduce from the proof a version of Rieffel's theorem.

In §3 we state several theorems which show the delicate nature of some of the characterizations of the RNP vs. similar characterizations of some other well-known classes of Banach spaces. It is hoped that the proof of the scalar case in §1 will at least make these results believable to the reader. The results given were selected for their beauty, they are often not the best results known.

One goal is to indicate how the area of research has mushroomed since the fundamental results of Pettis and others in the late 1930's. The Diestel-Uhl book [7] is recommended for additional information.

The remainder of this section is devoted to the definition of the RNP and a few preliminary remarks.

Let $([0,1],\Sigma,\lambda)$ denote the unit interval with its σ-algebra Σ of Borel sets and Lebesgue measure λ. All Banach spaces will be over the real numbers $\mathbb{R}$.

DEFINITION. A Banach space X has the <u>Radon-Nikodym Property</u> (RNP) if and only if whenever $\mu : \Sigma \to X$ is an additive measure satisfying $\|\mu(\cdot)\| \leq \lambda(\cdot)$, there exists a bounded measurable function $g : [0,1] \to X$ such that $\mu(E) = \int_E g\,d\lambda$ (Bochner integral) for every E in Σ.

This definition is equivalent to much stronger formulations of the RNP. If g is not required to be bounded but only Bochner integrable, then $([0,1],\Sigma,\lambda)$ can be replaced by any σ-finite measure space, and μ by any countably additive, absolutely continuous X-valued measure of bounded variation. It is not such semi-formal equivalences of the definition which interests us here, but rather the implications of the RNP on the structure of the Banach space X.

A few facts about the RNP are almost immediate from the definition:

(1) The RNP is invariant under isomorphisms.

(2) The RNP is inherited by subspaces.

(3) The RNP is separably determined. (That is, if every separable subspace of X has the RNP, then X does also.)

Statement (1) is trivial. Statement (2) is because the g in the definition must take almost all its range in the closed linear span of the range of μ. Statement (3) holds since any μ as in the definition must have separable range. The true importance of these facts was first pointed out by J.J. Uhl [31]. We shall say that a space X is a U-space provided every separable subspace of X is isomorphic to a subspace of a separable dual space. Uhl [31] points out that perhaps the best statement of the Dunford-Pettis theorem is the following.

THEOREM 1 (N. Dunford-B.J. Pettis [8]). Every U-space has the RNP.

(For example, consider $\ell^1(\Gamma)$ for uncountable Γ. It is easily seen to be a U-space. However, it is not separable, and it contains separable subspaces which are not themselves dual spaces [18].) Of course reflexive spaces are U-spaces.

The two classical spaces c_0 and $L^1[0,1]$ fail to have the RNP, hence so does any space which contains an isomorphic copy of one of these spaces, e.g., $C[a,b]$, $C[a,b]^*$, ℓ^∞, and L^∞ all fail the RNP.

§2. RIEFFEL'S RADON-NIKODYM THEOREM.

We first consider the scalar case.

THEOREM (Radon-Nikodym). <u>The real line</u> $\mathbb{R}$ <u>has the</u> RNP.

The proof we give is a slight simplification of the scalar case of a proof due to M. Rieffel [29].

Let $\mu : \Sigma \to \mathbb{R}$ be additive and satisfy $|\mu(E)| \leq \lambda(E)$, all $E \in \Sigma$. The basic idea is that a derivative g should be given locally by

$$g(x) = \lim \frac{\mu(E)}{\lambda(E)}$$

where the limit is taken as the E's somehow shrink to the point x. This idea is carried out globally. For notation, given E in Σ, let

$$\Sigma^+(E) = \{F \in \Sigma : \lambda(F) > 0, \quad F \subset E\}$$

and

$$A_E = \{\frac{\mu(F)}{\lambda(F)} : F \in \Sigma^+(E)\}.$$

Note that A_E is contained in the interval $[-1,1]$. The first step of the proof is a key lemma.

LEMMA 1. <u>Given</u> $E \in \Sigma^+([0,1])$ <u>and</u> $\varepsilon > 0$, <u>there exists</u> $F \in \Sigma^+(E)$ <u>such that</u> $\operatorname{diam}(A_F) < \varepsilon$.

PROOF. Let $\alpha = \inf A_E$, and choose G in $\Sigma^+(E)$ such that

$$\alpha \leq \frac{\mu(G)}{\lambda(G)} \leq \alpha + \frac{\varepsilon}{4} .$$

If $\operatorname{diam} A_G < \varepsilon$ we are done. Otherwise there exists B in $\Sigma^+(G)$ with

(*) $$\left|\frac{\mu(B)}{\lambda(B)} - \frac{\mu(G)}{\lambda(G)}\right| > \frac{\varepsilon}{2}$$

hence

$$\frac{\mu(B)}{\lambda(B)} > \alpha + \frac{\varepsilon}{2} .$$

Let A be a (necessarily countable) maximal disjoint family of such sets; say $A = (B_n)_n$. Let $F = G\setminus(\bigcup_n B_n)$. By maximality of A, if H is in $\Sigma^+(F)$, then

$$\alpha \leq \frac{\mu(H)}{\lambda(H)} \leq \alpha + \frac{\varepsilon}{2}$$

and so $\operatorname{diam}(A_F) < \varepsilon$.

It remains only to show that $\lambda(F) > 0$. But if $\lambda(F) = 0$, then $\mu(G) = \sum_n \mu(B_n)$ and we have

(#) $$\frac{\mu(G)}{\lambda(G)} = \sum_n \frac{\lambda(B_n)}{\lambda(G)}\left(\frac{\mu(B_n)}{\lambda(B_n)}\right) .$$

Now $\frac{\lambda(B_n)}{\lambda(G)} \geqq 0$ and $\sum_n \frac{\lambda(B_n)}{\lambda(G)} = 1$, and this with (*) implies that the last quantity is greater than $\alpha + \frac{\varepsilon}{2}$, a contradiction. Hence $\lambda(F) > 0$. QED

LEMMA 2. <u>Given</u> $\varepsilon > 0$ <u>there exists a countable partition</u> $\pi = (E_m)_m$ <u>of</u> $[0,1]$ <u>such that for all</u> m, $\operatorname{diam}(A_{E_m}) < \varepsilon$.

The proof of the second lemma from the first is an easy maxmality argument. We now complete the proof that μ has a derivative g.

Using Lemma 2, choose a sequence $(\pi_n)_{n=1}^{\infty}$ of countable partitions of $[0,1]$ such that

$$E \in \pi_n \Rightarrow \operatorname{diam}(A_E) < \frac{1}{n} .$$

We may assume π_{n+1} refines π_n, each n. Let

$$g_n = \sum_{E \in \pi_n} \frac{\mu(E)}{\lambda(E)} \chi_E .$$

Then $(g_n)_{n=1}^{\infty}$ is uniformly Cauchy and so converges uniformly to some g. If $E \in \Sigma$, and if $\pi_n = (E_m)$, then

$$|\mu(E) - \int_E g_n d\lambda| = \left|\sum_m \left[\frac{\mu(E \cap E_m)}{\lambda(E \cap E_m)} - \frac{\mu(E_m)}{\lambda(E_m)}\right] \lambda(E \cap E_m)\right|$$

$$\leq \frac{1}{n} \Sigma\lambda(E \cap E_m)$$

$$= \frac{1}{n} \lambda(E).$$

Hence $\mu(E) = \lim \int_E g_n d\lambda = \int_E g d\lambda$. QED

We now ask how this proof can fail if $\mathbb{R}$ is replaced by a Banach space X. The only place trouble can enter is in Lemma 1. Looking closely at (*) and (#) we see that if Lemma 1 is false, then there exists some $\varepsilon > 0$ and E in $\Sigma^+([0,1])$ such that every G in $\Sigma^+(E)$ can be partitioned into a countable collection $(B_n)_n \subset \Sigma^+(G)$ such that

$$\left\|\frac{\mu(G)}{\lambda(G)} - \frac{\mu(B_n)}{\lambda(B_n)}\right\| > \varepsilon.$$

We have

$$\frac{\mu(G)}{\lambda(G)} = \sum_n \alpha_n \frac{\mu(B_n)}{\lambda(B_n)}$$

where $\alpha_n = \frac{\lambda(B_n)}{\lambda(G)} \geq 0$ and $\sum_n \alpha_n = 1$. Note that we can then similarly partition each B_n, and proceed indefinitely.

Therefore, if some μ fails to be representable, then we can construct what we shall call an infinite ε-<u>jungle</u> in the unit ball X. By an infinite

ε-jungle we mean the following. We start with some point x_0 and write it as a σ-convex combination

$$x_0 = \sum_{i=1}^{\infty} \alpha_i x_i \qquad (\alpha_i \geqq 0, \quad \sum_{i=1}^{\infty} \alpha_i = 1)$$

with $\|x_0 - x_i\| \geqq \varepsilon$, all $i = 1,2,\ldots$ $\{x_0\}$ is called the first level of the jungle, $(x_i)_{i=1}^{\infty}$ is called the second level. We then write each x_i as such a σ-convex combination

$$x_i = \sum_{j=1}^{\infty} \alpha_j^{(i)} x_j^{(i)}$$

with $\|x_i - x_j^{(i)}\| \geqq \varepsilon$, all i and j. Then $(x_j^{(i)})_{i,j}$ is called the third level. We continue indefinitely.

We have proved the following.

THEOREM 2 (Rieffel [29] (1967)). _If_ X _fails to have the RNP then for some_ $\varepsilon > 0$ _there is an infinite_ ε-_jungle in the unit ball of_ X.

H. Maynard [22] proved the converse of this theorem in 1972 (see Theorem _4_ below.)

From Theorem 2, we can deduce easily Clarkson's result.

THEOREM 3 (Clarkson [5] (1936)). _If_ X _is uniformly convex, then_ X _has the RNP_.

PROOF. Recall that X is UC if and only if for every $\varepsilon > 0$ there exists $\delta < 1$ such that

$$\|x\| \leqq 1, \quad \|y\| \leq 1, \quad \|x-y\| \geqq \varepsilon \Rightarrow \|\tfrac{x+y}{2}\| \leqq \delta .$$

Note that if $\bar{x}$ and $(x_i)_i$ are in the unit ball, and

$$\bar{x} = \sum_{i=1}^{\infty} \alpha_i x_i, \quad \alpha_i \geqq 0, \quad \sum_{i=1}^{\infty} \alpha_i = 1, \quad \text{and} \quad \|\bar{x}-x_i\| \geqq \varepsilon,$$

then

$$\|\bar{x}\| = \frac{1}{2} \| \sum_{i=1}^{\infty} \alpha_i (\bar{x}+x_i) \| \leqq \sum_{i=1}^{\infty} \alpha_i \|\frac{\bar{x}+x_i}{2}\| \leqq \delta.$$

Now if we had an ε-jungle in the unit ball of X with $n+1$ levels or more, then all the terms in the first two levels would necessarily have norm $\leqq \delta^n$, which is impossible if n is large enough. QED

The reader should now find Theorem 4 below reasonable. For that theorem we need a few more definitions.

By an ε-<u>bush</u> we mean an ε-jungle where at each stage the convex combinations are only finite; that is,

$$x_0 = \sum_{i=1}^{n_0} \alpha_i x_i, \qquad x_i = \sum_{j=1}^{n_i} \alpha_j^{(i)} x_j^{(i)}$$

for some finite n_i's, etc.

By an ε-<u>tree</u> we mean an ε-bush where at each stage the convex combination is simply the average of two points; that is,

$$x_0 = \frac{1}{2} x_1 + \frac{1}{2} x_2, \quad x_1 = \frac{1}{2} x_1^{(1)} + \frac{1}{2} x_2^{(1)} ,$$

etc.

A Banach space Y is said to be <u>finitely representable</u> in a Banach space X if for every finite dimensional subspace E of Y and every $\eta > 1$ there is an isomorphism $T : E \to X$ (into) such that $\|T\|\|T^{-1}\| < \eta$. If P is a property for Banach spaces, we say that a space X is <u>super</u> P provided

$$(Y \text{ finitely representable in } X) \Rightarrow (Y \text{ has property } P).$$

We say that X has the _Krein-Milman Property_ (KMP) provided every closed bounded convex subset K of X has an extreme point. (It is then automatic that each such K is the closed convex hull of its extreme points [19].)

3. SOME CHARACTERIZATIONS OF THE RNP.

THEOREM 4 (The RNP vs. uniform convexity). _Let X be a Banach space._

(I) (R.C. James-P.Enflo [11]). _Each of the following statements is equivalent to X being isomorphic to a uniformly convex space._

(a) _For every_ $\varepsilon > 0$ _there exists an integer_ n _such that every_ ε-_jungle in the unit ball of_ X _has at most_ n _levels._

(b) _Same as_ (a) _with_ ε-_bushes._

(c) _Same as_ (a) _with_ ε-_trees._

(II) (M. Rieffel-H. Maynard (see [7])). _Each of the following statements is equivalent to_ X _having the_ RNP.

(a′) _For every_ $\varepsilon > 0$ _the unit ball of_ X _does not contain an infinite_ ε-_jungle._

(b′) _Same as_ (a′) _with_ ε-_bushes._

(II′) (C. Stegall [30]). _If_ X _is a dual space, then the_ RNP, (a′) _and_ (b′) _are equivalent to_

(c′) _Same as_ (a′) _with_ ε-_trees._

(II″) (J. Bourgain [3]). _There exists Banach spaces which satisfy_ (c′) _but which do not have the_ RNP.

(III) _Each of the following statements is equivalent to_ X _being isomorphic to a uniformly convex space._

(i) (R.C. James-P. Enflo [11]). X _is super-reflexive._

(ii) (G. Pisier [27]). X _is super_-RNP.

(iii) (G. Pisier [27]). X _is super_-KMP.

For the next theorem we need several additional definitions.

We say that a point x in a set A is an _extreme point_ of A provided x cannot be written as a convex combination of other points of A. (Note that A need not be convex in this definition). A functional f in the dual X* of X is said to _support_ a set A at a point x in A provided $f(x) = \sup f(A)$; we let $S_A = \{f \in X^* : f$ supports $A\}$. We say that f _strongly exposes_ A at a point x in A provided f supports A at x and provided

$$(x_n)_1^\infty \subset A, \quad f(x_n) \to f(x) \Rightarrow x_n \to x.$$

In this case, the x is unique and is necessarily an extreme point of A. We let

$$E_A = \{f \in X^* : f \text{ strongly exposes } A\}.$$

Recall that a space X is reflexive iff its unit ball is weakly compact iff every bounded weakly closed set is weakly compact.

For the next two theorems, the reader should consider the role of α in the proof of Lemma 1 in §2. In these two theorems the placement of the word "convex" is critical. Where it appears in parentheses it may be removed and an equivalent statement results; otherwise its presence or lack of presence is usually critical.

THEOREM 5 (The RNP vs. reflexivity).

(I) (E. Bishop-R.R. Phelps [1]). _Every Banach space_ X _has the following property: for every closed bounded convex set_ $K \subset X$, _the set_ S_K _is dense in_ X*.

(II) (R.C. James (see [6])). *A Banach space* X *is reflexive if and only if and only if for every weakly closed bounded (convex) set* $K \subset X$, *the set* S_K *is all of* X^*.

(III) *Each of the following statements is equivalent to the* RNP *for a Banach space* X.

(a) (R.R. Phelps [26]). *For every closed bounded (convex) set* A, E_A *is non-void (and then* E_A *is automatically a norm dense* G_δ *in* X^* (R. Huff-P.D. Morris [16]).)

(b) (J. Bourgain-C. Stegall (see [28])). *For every closed bounded convex set* $K \subset X$, *the set* S_K *is of second category in* X^*.

(c) (R. Huff-P.D. Morris [16]). *For every norm closed bounded subset* A *of* X, *the set* S_A *is non-void (and then* S_A *is automatically of second category in* X^*.)

(d) (J. Bourgain [28]). *Same as* (c) *with weakly closed bounded sets.*

(e) (R. Huff-P.D. Morris [16]). *If* A *is closed and* U *is a bounded open set with* $A \subset U \subset X$, *then* $\overline{co}(A) \neq \overline{co}(U)$.

We remark that there exist characterizations of the RNP similar to those in (III), but with operators replacing functionals; see Bourgain [2] (also [14]).

The above results already establish a connection between the RNP and the KMP. In 1973, before the above results were known, J. Diestel conjectured that the two properties were equivalent. The following results have been established.

THEOREM 6 (The RNP vs. the KMP). *Let* X *be a Banach space.*

(I) (J. Lindenstrauss (see [26])). *If* X *has the* RNP, *then* X *has the* KMP. *(In fact* (R.R. Phelps [26]), *in this case every closed bounded convex set is the closed convex hull of its strongly exposed points.)*

(II) (a) (R. Huff-P.D. Morris [16]). If every norm closed bounded subset of X has an extreme point, then X has the RNP.

(b) (J. Bourgain [33]). Same as (a) with weakly closed bounded sets.

(c) (R. Huff-P.D. Morris [15]). If X is a dual space, and if X has the KMP, then X has the RNP.

We remark that G.A. Edgar [9], [10] has established versions of Choquet's theorem in spaces with the RNP. (See also Mankiewicz [21]).

In view of the above, Diestel's conjecture reduces to the following question. If every closed bounded convex subset of X has an extreme point, must every (weakly) closed bounded set have one? It should be remarked that there exist closed bounded sets A without extreme points, but with $\overline{co}(A)$ having many extreme points [16].

The proofs of the statements in Theorem 6 (II) are indirect: assuming the failure of the RNP, one constructs sets without extreme points. More direct proofs might be enlightening.

In §1 we defined U-spaces (see Theorem 1). In [31], Uhl asked if every space with the RNP is necessarily a U-space. This question was not completely answered until 1978. The following results are known.

THEOREM 7 (The RNP vs. U-space).

(I) (N. Dunford-B.J. Pettis (1940)) Every U-space has the RNP.

(II) (J. Bourgain and F. Delbaen [4]; independently, P.W. McCartney-R.C. O'Brien [23], both 1978). There exist Banach spaces which have the RNP and are not U-spaces.

(III) (C. Stegall [30] (1975)). If X is a dual space with the RNP, then it is a U-space. (In fact, a dual Y* has the RNP if and only if every separable subspace of Y has a separable dual.)

Call a space X an (Nℓ)-space if no subspace of X is isomorphic to ℓ^1.

For many years it was conjectured that a separable space was $(N\ell)$ if and only if it had a separable dual. In view of Stegall's result above, an equivalent conjecture would be that X^* has the RNP if and only if X is an $(N\ell)$-space. This conjecture was proved false in 1974 by R.C. James [17]. Nevertheless, the conjecture is fascinatingly close to being true. The following results hold; see [12] for additional results.

THEOREM 8 (RNP dual spaces vs. duals of $(N\ell)$-spaces). *Let* X *be a dual space, say* $X = Y^*$.

(I) (Krein-Milman). *Every weak* closed, bounded convex set in* X *is the weak* closed convex hull of its extreme points.*

(II) (E. Odell-H.P. Rosenthal-R. Haydon [12]). Y *is an* $(N\ell)$-*space if and only if every weak* closed, bounded convex set in* X *is the norm closed convex hull of its extreme points.*

(III) (R. Huff-P.D. Morris [15]). X *has the* RNP *if and only if every norm closed, bounded convex set in* X *is the norm closed convex hull of its extreme points.*

(IV) (R.C. James-J. Lindenstrauss-C. Stegall [20]). *There exists a space* X *such that all of its successive duals* $X, X^*, X^{**}, X^{***}, \ldots, X^{(n)}$ *are* $(N\ell)$-*spaces, every even dual* $X^{(2n)}$ *fails to have the* RNP, *and every odd dual* $X^{(2n+1)}$ *has the* RNP.

The above theorems give but a sampling of the theory about the RNP. Many open problems exist (see [7] for a start). For example, an important open problem is to characterize basic sequences which generate RNP-spaces. It is known that a boundedly complete basic sequence does, but that condition is too strong. There ought to be an analogue for RNP for the following result [32]: if X has a basis, then X is reflexive if and only if every basis for X is boundedly complete.

REFERENCES

1. E. Bishop and R.R. Phelps, The support functionals of a convex set, Proc. Sympos. Pure Math., vol. 7, Amer. Math. Soc., Providence, R.I. (1963), 27-35.

2. J. Bourgain, On dentability and the Bishop-Phelps property, Israel J. Math. 28 (1977), 265-271.

3. __________, preprint.

4. __________ and F. Delbaen, A special class of L_∞ spaces, Acta Math., to appear.

5. J.A. Clarson, Uniformly convex spaces, Trans. Amer. Math. Soc. 40 (1936), 396-414.

6. J. Diestel, Geometry of Banach Spaces - Selected topics, Lecture Notes in Math., vol. 485, Springer-Verlag, Berlin and New York.

7. __________ and J.J. Uhl, Jr., Vector Measures, Amer. Math. Soc. Math. Surveys, No. 15, Providence, R.I. (1977).

8. N. Dunford and B.J. Pettis, Linear operators on summable functions, Trans. Amer. Math. Soc. 47 (1940), 323-392.

9. G.A. Edgar, A noncompact Choquet Theorem, Proc. Amer. Math. Soc. 49 (1975), 354-358.

10. __________, Extremal integral representations, J. Functional Anal. 23 (1976), 145-161.

11. P. Enflo, Banach spaces which can be given an equivalent uniformly convex norm, Israel J. Math. 13 (1973), 281-288.

12. R. Haydon, Some more characterizations of Banach spaces containing ℓ_1, Math. Proc. Camb. Phil. Soc. 80 (1976), 269-276.

13. R. Huff, Dentability and the Radon-Nikodym property, Duke Math. J. 41 (1974), 111-114.

14. _______, On non-density of norm-attaining operators, Revue Roum. Math. Pures Appl. (to appear).

15. _______ and P.D. Morris, Dual spaces with the Krein-Milman property have the Radon-Nikodym property, Proc. Amer. Math. Soc. 49 (1975), 104-108.

16. _______ and __________, Geometric characterizations of the Radon-Nikodym property in Banach spaces, Studia Math. 61 (1976), 157-164.

17. R.C. James, A separable somewhat reflexive Banach space with nonseparable dual, Bull. Amer. Math. Soc. 80 (1974), 738-743.

18. J. Lindenstrauss, On a certain subspace of ℓ_1, Bull. Acad. Polon. Sci. Ser. Sci. Math. Astronom. Phys. 12 (1964), 539-542.

19. _______________, On extreme points in ℓ_1, Israel J. Math. 4 (1966), 59-61.

20. _______________ and C. Stegall, Examples of separable spaces which do not contain ℓ_1 and whose duals are non-separable, Studia Math. 56 (1975), 81-105.

21. P. Mankiewicz, A remark on Edgar's extremal integral representation theorem, Studia Math. 68 (1978), 259-265.

22. H. Maynard, A geometric characterization of Banach spaces having the Radon-Nikodym property, Trans. Amer. Math. Soc. 185 (1973), 493-500.

23. P.W. McCartney and R.C. O'Brien, A separable Banach space with the Radon-Nikodym property which is not isomorphic to a subspace of a separable dual, (preprint).

24. D.P. Milman, On some criteria for the regularity of spaces of type (B), Dokl. Aka. Nauk. SSSR 20 (1938), 243-246. (Russian)

25. B.J. Pettis, A proof that every uniformly convex space is reflexive, Duke Math. J. 5 (1939), 249-253.

26. R.R. Phelps, Dentability and extreme points in Banach spaces, J. Functional Analysis 16 (1974), 78-90.

27. G. Pisier, Martingales with values in uniformly convex spaces, Israel J. Math. 20 (1975), 326-350.

28. J. Rainwater seminar notes, Univ. of Washington, Seattle, 1976-77.

29. M. Rieffel, Dentable subsets of Banach spaces, with application to a Radon-Nikodym theorem, Functional Analysis (Proc. Conf., Irvine, Calif., 1966), (B.R. Gelbaum, editor), Academic Press, London, Thompson, Wash., D.C., (1967), pp. 71-77.

30. C. Stegall, The Radon-Nikodym property in conjugate Banach spaces, Trans. Amer. Math. Soc. 206 (1975), 213-223.

31. J.J. Uhl, Jr., A note on the Radon-Nikodym property for Banach spaces, Rev. Roum. Math. Pures Appl. 17 (1972), 113-115.

32. M. Zippin, A remark on bases and reflexivity in Banach spaces, Israel J. Math. 6 (1968), 74-79.

33. J. Bourgain, A geometric characterization of the Radon-Nikodym property in Banach spaces, Compositio Math. 36 (1978), 3-6.

Pennsylvania State University

Contemporary Mathematics
Volume 2
1980

THE ORLICZ-PETTIS THEOREM

N. J. Kalton*

1. Introduction

It is my aim in this talk to give a summary of certain aspects of the development of the Orlicz-Pettis theorem since its origins fifty years ago. During the course of its evolution this theorem has evolved almost beyond recognition, and the techniques developed for its study have themselves helped to illuminate a number of ideas in functional analysis.

This will be only a very partial survey; in such a short time it would be impossible to cover adequately all the directions taken by recent research. It will reflect also my personal interests in the area. The Orlicz-Pettis theorem has been an important catalyst in the development of the theory of (non-locally convex) F-spaces.

2. The basic Orlicz-Pettis theorem

Let us start by giving two equivalent formulations of the classical Orlicz-Pettis theorem.

Theorem 1 (a) Let X be a Banach space and let Σx_n be series in X with the property that for every subseries Σx_{n_k} there exists $x \in X$, with

$$x^*(x) = \sum_{k=1}^{\infty} x^*(x_{n_k}) \qquad x^* \in X^* .$$

*The author was partially supported during the preparation of this paper by NSF grant MCS-7903079.

Copyright © 1980, American Mathematical Society

Then Σx_n converges in X [Briefly: a weakly subseries convergent series is convergent].

(b) Let X be a Banach space, and let $\mathcal{A}$ be a σ-algebra of sets. Suppose $\mu\colon \mathcal{A} \to X$ is set function such that $x^* \circ \mu$ is a countably additive measure for all $x^* \in X^*$. Then μ is a countably additive vector measure. [Briefly: a weak countably additive measure is countably additive for the norm].

With the restriction that X is weakly sequentially complete, Theorem 1 (a) was proved by Orlicz in 1929 [22]. However 1 (a) was known to Orlicz without this restriction (see [2]). The first accessible proof was given by Pettis in 1938 ([24]), and it was Pettis who pointed out the applications of the result to vector measures, in particular 1 (b).

The proof of the theorem by both Orlicz and Pettis depends on a special property of the Banach space ℓ_1, now known as the Schur property. Schur had proved a result for summability methods which translates to the statement that if $\{x_n\}$ is a weakly convergent sequence in ℓ_1 then $\{x_n\}$ converges in norm. If we identify $\ell_1^* = \ell_\infty$, then in fact it is necessary only to check that $\lim_{n\to\infty} x^*(x_n)$ exists for all $x^* \in m_0$ where $m_0 \subset \ell_\infty$ is the subspace of finitely-valued sequences.

Let us now sketch a proof of Theorem 1 (a). First observe that we can assume X separable, and next can reduce the problem to showing the impossibility of

$$\Sigma x_n \text{ weakly subseries convergent}$$

$$||x_n|| \geq 1 \qquad n \in N .$$

Pick $x_n^* \in X^*$ with $||x_n^*|| \leq 1$ and

$$x_n^*(x_n) = 1 \qquad n \in N .$$

Since X is separable, we can pass to a subsequence and suppose for some

$x_0^* \in X^*$

$$\lim_{n\to\infty} x_n^*(x) = x_0^*(x) \qquad x \in X .$$

Suppose $t = \{t_n\}_{n=1}^{\infty} \in m_0$; then for some $y_t \in X$

$$x^*(y_t) = \sum_{n=1}^{\infty} t_n x^*(x_n) \qquad x^* \in X^*$$

and so

$$\lim_{m\to\infty} \sum_{n=1}^{\infty} t_n(x_m^*(x_n) - x_0^*(x_n)) = 0 .$$

It is easy to check that for each m

$$\sum_{n=1}^{\infty} |x_m^*(x_n) - x_0^*(x_n)| < \infty$$

and so by the Schur property of ℓ_1

$$\lim_{m\to\infty} \sum_{n=1}^{\infty} |x_m^*(x_n) - x_0^*(x_n)| = 0$$

In particular

$$\lim_{n\to\infty} |x_n^*(x_n) - x_0^*(x_n)| = 0$$

and $x_0^*(x_n) \to 1$ contradicting the convergence of $\Sigma x_0^*(x_n)$.

A later rather more brutal Banach space style proof was given by Bessaga and Pelczynski in 1958 ([4]), exploiting the properties of basic sequences. Still more recently Uhl [29] has given another proof based on the Pettis-measurability theorem, showing a rather curious tie-up between these two at first sight unrelated results.

3. Further developments in the context of Banach spaces

As far as Banach spaces are concerned the main thrust of research into improving the Orlicz-Pettis theorem has been to attempt to replace the assumption in 1 (b) that $x^* \circ \mu$ is countably additive for all $x^* \in X^*$ by some smaller collection of linear functionals. The definitive result here is due to Diestel and Faires (1974, [7]) and it perhaps suggests that the role of the weak topology in the Orlicz-Pettis theorem is exaggerated.

A simple example shows that we cannot simply demand that $x^* \circ \mu$ is countably additive for a total subspace of X^*. Let $\mathcal{P} N$ be the power set of N and let $\mu: \mathcal{P} N \to \ell_\infty$ be defined by

$$\mu(A) = \chi_A$$

(the characteristic function of A). Then $x^* \circ \mu$ is countably additive if $x^* \in \ell_1 \subset (\ell_\infty)^*$, but μ is not countably additive. The Diestel-Faires theorem shows that this is the "only" such example.

Theorem 2 Let X be a Banach space containing no copy of ℓ_∞, and let $\mathcal{A}$ be a σ-algebra of sets. Let $\mu: \mathcal{A} \to X$ have the properties that $x^* \circ \mu$ is countably additive for x^* in some total subspace H of X^*. Then μ is countably additive.

Theorem 2 depends on two deep results.

Theorem 3 (Grothendieck [14]) m_0 is a barrelled subspace of ℓ_∞; equivalently if $(\phi_i : i \in I)$ is any collection of finitely additive bounded set functions on $\mathcal{P} N$ with

$$\sup_{i \in I} |\phi_i(A)| < \infty \qquad A \in \mathcal{P} N$$

then

$$\sup_{A \in \mathcal{P}N} \sup_{i \in I} |\phi_i(A)| < \infty$$

<u>Theorem 4</u> (Rosenthal [26]) <u>If</u> X <u>is a Banach space not containing</u> ℓ_∞ <u>and</u> $T: \ell_\infty \to X$ <u>is a bounded linear operator, then</u> T <u>is weakly compact</u>.

Now a proof of the Diestel-Faires theorem can be built as follows. We can suppose $\mathcal{A} = \mathcal{P}\,\mathbb{N}$; μ induces a linear map $T_0: m_0 \to X$. The hypotheses make T_0 a linear map with closed graph; but now Theorem 3 allows us to use the Closed Graph Theorem, since m_0 is a barrelled normed space. Hence T_0 is bounded and induces a bounded linear extension $T: \ell_\infty \to X$. Then by Theorem 4, T is weakly compact. From here it is an easy step to show that $x^* \circ \mu$ is countably additive for every $x^* \in X^*$ and then apply the original Orlicz-Pettis theorem.

The appearance of the Closed Graph Theorem is significant here, as we shall see later when we leave the secure surrounds of Banach spaces.

4. Locally convex spaces

The extension of the Orlicz-Pettis theorem to generally locally convex spaces was achieved by McArthur in 1967 [21], after earlier results by Grothendieck [15]; another proof was given by Robertson [25].

<u>Theorem 5</u> <u>Let</u> X <u>be a locally convex topological vector space and let</u> $\mathcal{A}$ <u>be a σ-algebra of sets</u>. <u>Suppose</u> $\mu: \mathcal{A} \to X$ <u>is a weakly countably additive measure. Then</u> μ <u>is countably additive</u>.

Subsequent work in this area has largely been directed to pushing countable additivity further than the original topology on X; see for example [3]. For some recent developments see Dierolf [5] and Graves [12].

5. Abelian groups

We have already seen an interaction between the Closed Graph theorem and the Orlicz-Pettis theorem. In fact, Theorem 5 can be understood as a closed graph-type result: the identity map from X with the weak topology into X with original topology has closed graph and Theorem 5 answers its "continuity" at least on subseries convergent series.

In 1970, I became interested in this link after establishing a mild variation of Ptak's Closed Graph Theorem. In [16], I showed that if E is a Mackey space whose dual E^* is weak* sequentially complete and F is a separable Frechet space then every linear map $T: E \to F$ with closed graph is continuous. The hypotheses on E are satisfied if we take $E = m_0$ with the Mackey topology induced by ℓ_1. Then the theorem can be used to obtain a weaker form of the Diestel-Faires theorem (Theorem 2), valid for separable Banach spaces.

At the same time, Stiles [27] established the Orlicz-Pettis theorem (Theorem 1) for F-spaces (complete metric linear spaces) with a Schauder basis. Of course, the basis assures the non-triviality of the dual space, but there is no longer any close relationship between the weak topology and the metric topology on the space. In this sense, Stiles's result was a very important and significant departure from earlier results. On seeing Stiles's paper, I was prompted to examine whether some form of Orlicz-Pettis theorem could be proved in F-spaces using the Closed Graph Theorem.

In fact, Polish abelian groups proved to be the appropriate setting. The main step was to establish the suitability as a domain space in the Closed Graph Theorem for groups of the integer-valued analogue of m_0 with its Mackey topology induced by ℓ_1. The main result was ([17]).

<u>Theorem 6</u> <u>Let</u> G <u>be an abelian Polish group and let</u> α <u>be any weaker Hausdorff group topology on</u> G. <u>Suppose</u> $\mu: A \to G$ <u>is</u> α<u>-countably additive</u>. <u>Then</u> μ <u>is countably additive</u>.

The shortest and neatest proof of Theorem 6 is due to Drewnowski ([8], see also [10]). His approach was to establish first

<u>Theorem 7</u> <u>Let</u> (G, β) <u>be a separable abelian topological group and let</u> α <u>be a weaker Hausdorff group topology on</u> G. <u>Suppose</u> β <u>has a base of</u> α<u>-closed neighborhoods of</u> 0. <u>Then any</u> α<u>-countably additive measure is</u> β<u>-countably additive</u>.

The proof of Theorem 7 is a simple argument based on the Baire Category Theorem. Now Theorem 6 is proved by taking γ to be the maximal group topology on G weaker than the original metric topology for which μ is countably additive. Define γ^* to be that topology whose base at 0 consists of all γ-closed metric neighborhoods of 0. Use Theorem 7 to deduce $\gamma = \gamma^*$ and the basic Closed Graph Theorem for groups to deduce that γ is the metric topology (cf. [18] p. 213). In fact a version of Theorem 6 may be proved for non-abelian groups ([8]).

In 1973 Anderssen and Christensen [1] extended Theorem 6 to groups with an analytic topology (i.e. abelian topological groups which are the continuous images of Polish spaces). They showed it is necessary only to have the identity from (G, α) into G a Borel map, and this is automatic if G has an analytic topology. Continuing this line, recently Labuda [20], Pachl [23] and Graves [13] have established:

<u>Theorem 8</u> <u>(Graves-Labuda-Pachl) Let</u> G <u>be a complete abelian topological group and let</u> α <u>be a Hausdorff group topology on</u> G <u>weaker than the original so that the identity</u> $i: (G, \alpha) \to G$ <u>is universally measurable</u>. <u>Then any α-countably additive G-valued measure is countably additive</u>.

It is worth briefly sketching the ideas of the proof as given by Graves [13]. It can be shown to be sufficient to prove the result when i is universally Lusin-measurable, and for a measure μ on $\mathcal{P}\,\mathbb{N}$. If we topologize $\mathcal{P}\,\mathbb{N}$ as $\{0,1\}^\omega$ it is compact metric, and considered as the Cantor group it has a Haar measure λ say. Now $\mu: \mathcal{P}\,\mathbb{N} \to G$ is universally Lusin measurable. It will suffice (since G is complete) to show that $\mu\{n\} = e_n \to 0$; and we may again reduce to the case where $e_n \notin V$ for some V of 0, and prove the result by contradiction.

Now choose $\mathcal{M} \subset \mathcal{P}\,\mathbb{N}$ to be compact such that $\lambda(\mathcal{M}) > \frac{1}{2}$ and $\mu|\mathcal{M}$ is continuous. For each n, let $\mathcal{M}_n^- = \{A \in \mathcal{M}, n \notin A\}$ and $\mathcal{M}_n^+ = \{A: n \notin A, A \cup \{n\} \in \mathcal{M}\}$. Then

$$\begin{aligned}\lambda(\mathcal{M}_n^-) + \lambda(\mathcal{M}_n^+) &= \lambda(\mathcal{M}) \\ &> \frac{1}{2} \\ &= \lambda\{A: n \notin A\}\end{aligned}$$

Hence there exists $A_n \in M_n^- \cup M_n^+$. In particular $A_n \in M$ and so by passing to a subsequence we may suppose $A_n \to A_\infty \in M$ and $A_n \cup \{n\} \to A_\infty \in M$. Thus

$$e_n = \mu\{n\} = \mu(A_n \cup \{n\}) - \mu(A_n)$$

$$\to 0 .$$

This contradiction is sufficient to prove the result.

It seems to me that Theorem 8 is probably the final word on this line of development.

6. F-spaces

Interesting problems arise in the attempt to extend the Diestel-Faires Theorem (Theorem 2) to general non-locally convex F-spaces. One might hope that if X is an F-space containing no copy of ℓ_∞ and $\mu: A \to X$ is a countably additive measure for some weaker Hausdorff vector topology then μ is countably additive. Recently this reasonable hope was exploded by Turpin [28] who showed it to be false when X is a certain Orlicz sequence space $\ell_\phi(I)$ over an uncountable index set I(it is of course true if I is countable by Theorem 6).

If we analyze the breakdown of Theorem 2, we find that an analogue of Theorem 4 is true in this setting; an important extension of Rosenthal's result was obtained by Drewnowski ([9] and [11]) and this would suffice for the argument. In place of Theorem 3 however one would require m_0 ultrabarrelled and this is false ([6]). Thus we find that Theorem 2 fails in this setting because it is not possible to show that μ is bounded and then use operators on ℓ_∞. This is exactly what happens in Turpin's example.

Turpin does raise the question of what happens in the case when X is locally bounded. This again reduces to a question concerning the space m_0: is it "p-barrelled" for $0 < p < 1$.

In the context of groups, Labuda [19] has proved substitutes for the Diestel-Faires theorem involving the idea of copies of rings of sets.

References

1. N. J. M. Anderssen and J. P. R. Christensen, Some results on Borel structures with applications to subseries convergence in abelian topological groups, Israel J. Math. 15(1973) 414-420.

2. S. Banach, Théorie des Opérations Linéaires, Warsaw, 1932.

3. G. Bennett and N. J. Kalton, FK-spaces containing c_0, Duke Math. J. 39(1972) 561-582.

4. C. Bessaga and A. Pelczyński, On bases and unconditional convergence of series in Banach spaces, Studia Math. 17(1958) 151-164.

5. P. Dierolf, Theorems of the Orlicz-Pettis type for locally convex spaces, Mann. Math 20(1977) 73-94.

6. P. Dierolf, S. Dierolf and L. Drewnowski, Remarks and examples concerning unordered Baire-like and ultrabarrelled spaces, to appear.

7. J. Diestel and B. Faires, On rector measures, Trans. Amer. Math. Soc. 198(1974) 253-271.

8. L. Drewnowski, On the Orlicz-Pettis type theorems of Kalton, Bull. Acad. Polon. Sci. 21(1973) 515-518.

9. L. Drewnowski, Un théoréone sur les operators de $\ell_\infty(\Gamma)$, Comptes Rendus Acad. Sci. (Paris) Series A 281(1975) 967-969.

10. L. Drewnowski, Another note on Kalton's theorems, Studia Math. 52(1975) 233-237.

11. L. Drewnowski, An extension of a theorem of Rosenthal on operators acting from $\ell_\infty(\Gamma)$, Studia Math. 57(1976) 209-215.

12. W. H. Graves, A topological linearization of vector measures, Carolina Lecture Series 6, 1976.

13. W. H. Graves, Universal Lusin measurability and subfamily summable families in abelian topological groups, Proc. Amer. Math. Soc. 235(1979) 45-50.

14. A. Grothendieck, Espaces Vectoriels Topologiques, Sao Paulo, 1954.

15. A. Grothendieck, Sur les applications linéaires faiblement compactes d'espaces du type C(K), Canad. J. Math. 5(1953) 129-173.

16. N. J. Kalton, Some forms of the closed graph theorem, Proc. Cambridge Philos. Soc. 70(1971) 401-408.

17. N. J. Kalton, Subseries convergence in topological groups and vector measures, Israel J. Math. 10(1971) 402-412.

18. J. L. Kelley, General Topology, van Nostrand, 1955.

19. I. Labuda, Discrete copies of rings of sets in groups and Orlicz-Pettis theorems, Canad. J. Math. 30(1978) 748-755.

20. I. Labuda, Universal measurability and summable families in trs, Proc. Kon. Ned. Akad. von Welersch A 82(1979) 27-34.

21. C. W. McArthur, On a theorem of Orlicz and Pettis, Pacific J. Math. 22(1967) 297-302.

22. W. Orlicz, Beitrage zur Theorie des Orthogonalent wicklungen II, Studia Math. 1(1929) 241-255.

23. J. Pachl, A note on the Orlicz-Pettis theorem, Proc. Kon. Ned. Akad. von Werensch A 82(1979) 35-37.

24. B. J. Pettis, On integration in vector spaces, Trans. Amer. Math. Soc. 44(1938) 277-304.

25. A. P. Robertson, On unconditional convergence in topological vector spaces, Proc. Roy. Soc. Edinburgh Sect. A 68(1969) 145-157.

26. H. P. Rosenthal, On relatively disjoint families of measures with some applications to Banach space theory, Studia Math. 37(1970) 13-36.

27. W. J. Stiles, On subseries convergence in F-spaces, Israel J. Math. 7(1970) 53-56.

28. P. Turpin, Properties of Orlicz-Pettis or Nikodym type and barrelledness conditions, Ann. Inst. Fourier (Grenoble) 28 (3) (1978) 67-86.

29. J. J. Uhl, this volume.

Department of Mathematics
University of Missouri-Columbia
Columbia, Missouri 65211

Contemporary Mathematics
Volume 2
1980

APPLICATIONS OF VECTOR MEASURES

Igor Kluvánek

The theme of this essay is, partially, the result of a misunderstanding and of unwarranted courage. I interpreted the suggestion to speak about an application of vector measures as an invitation to discuss applications of vector measures. Maybe Pettis's presence (it happened in late 1978) inspired confidence and so I accepted the challenge. I believed, and still believe, that it would be a useful thing to have available a reliable survey of the basic applications of vector measures. In fact, I feel that my exposition will serve a good purpose if I convince my audience, and readers, about the desirability of such a survey to the extent that other mathematicians will improve on my efforts and compose more objective, complete and scholarly surveys on the role vector measures have played and, more importantly, are destined to play in mathematics and beyond. So much for the explanation and excuse of my boldness in the choice of the theme.

Another excuse and explanation concerns the style. Since I do not wish to pretend that I am presenting here anything more than my personal views and, at the same time, I do wish to incorporate some improvements on them which resulted from my encounters with the members of the audience after delivering the talk, I maintain, in the written version, the informal style of the spoken form. This liberty in no way represents a lack of regard for the discussed subject and, much less so, a disrespect to the reader. By taking it I mean rather the opposite. I hasten to add that, in the written version, I treat the subject more thoroughly and include

Copyright © 1980, American Mathematical Society

cases and technicalities which I did not consider suitable for the one-hour presentation.

The whole essay consists of three (unequal) parts. In the first one, I make some remarks, often personal, of a general character, concerning the situation in applications of vector measures.

The second part represents an attempt to draw up a list of the principal fields of applications of vector valued measures. Some attention is payed to the history of the subject. But still, it is very little more than a sort of a catalogue supplemented by a few references. Needless to say, the references are far from complete. Given my limitations, it is impossible to cover everything. As will be apparent, all the worth-while results about vector measures are, in my opinion, related to applications of one sort or another. On the other hand, I believe that, by following the references on any particular theme, the reader would be able to reconstruct a fairly complete picture.

In the third part, a few examples are worked out in, I hope, sufficient detail. The choice of these examples is dictated, first and foremost, by my being at least partially acquainted with them and, secondly, by the opportunity they provide to make some points about the theme of this essay. These examples suggest themes of investigation of classical origins but, from the vector measure theory point of view, developed to a much lesser extent than those mentioned in the second part.

1. General remarks

I wish to suggest that the theory of vector measures is basically related to the very core of analysis or, to be conservative, at least of functional analysis and, hence, to its most important applications. Consequently, to speak about applications means to speak about the sources and origins of vector measures. As in other vital branches of mathematics, we may observe a seemingly paradoxical situation, namely that many applications of vector measures occurred before, even long before, the formulation of the central concepts and theorems of the theory itself. It is my intention to support this suggestion by noting some cases of implicit use of vector measure techniques which lead, or may have lead, to the creation of the theory of vector measures as an "independent" discipline.

If some plausibility is granted to my suggestion, then, naturally the question arises, why is it that vector measures are not studied more intensively and why is the theory of vector measures not known more widely? In spite of the increasing number and status of workers in the field, there can still be felt some lack of enthusiasm for or even a suspicion of vector measures. The suggestion is often made, not necessarily explicitly, that they are an oddity studied only *par default* in order to gain degrees and honours.

Not so long ago, at a conference, I was sitting next to a known mathematical physicist. During the usual small-talk I confessed that my mathematical denomination had to do with vector measures. He was quite astonished that anybody could consider vector measures a worth-while enterprise for a grown-up person: "Can one do anything with vector measures that can't be done with ordinary measures? Anyway isn't the whole thing a trivial hoax?" Many of my comrades-in-arms may have had similar experiences and, hence, would be inclined to agree with me. My point is that I am not simply inventing a heresy to refute easily and so to cover myself with the laurels of victory. By the way, the incident with the mathematical physicist took a twist of irony when I discovered, much later of course, that the vector measure approach represented the key to some of his problems.

I would like to offer a few explanations for this attitude towards vector measure theory and its applications.

The possibility of applications of results of one branch of a science in another one, or of results of a science in another science, presupposes an internal unity or interdependence among various sciences. I believe, because I am speaking about applications of vector measures if for no other reasons, that such a unity and interdependence really exists. I stress that this is rather a statement of belief or philosophical allegiance than a mathematical theorem, even if many mathematical theorems would not be possible, or would loose their meaning and significance, if this belief were not true. However, in the aftermath of positivistic thought, the belief in the unity of the World and, consequently, in the unity of sciences, is not generally held or, at least, often it is not put into practice. There are, of course, notable counter-examples. But it would be impossible to explain otherwise the prevalence of convention. It would be impossible to explain otherwise the spasmodic way sciences, including mathematics, develop.

It is unbelievable that a mathematician of such standing as G.H. Hardy could have prided himself that he had never invented anything useful, to rephrase in plain Slovak the leading idea of his *A Mathematician's Apology*. If I may misquote A. Einstein [43, p. 49], "This is an interesting example of the fact that even scholars of audacious spirit and fine instinct can be obstructed in the interpretation of facts by philosophical prejudices". I can't offer any other explanation of how it could be possible that such a great mathematician could ignore the implications of his work for the rest of mathematics, applied or otherwise.

In this atmosphere, the conventions and collective or individual subjective opinions acquire undue importance. Hence, some themes of research come inexplicably into vogue and, sooner or later, inexplicably wither away.

The chances are that a theme of investigation just "dreamed up" without a connection with the main stream of mathematics or with serious problems from outside, will be superficial, even trivial. If one does not have an aim to achieve and objective criteria of whether the aim is achieved, then there is no reason why one should not introduce assumptions to make all solutions easy; there is no reason not to define all difficulties away. That the results then obtained are useless, for any other purpose but obtaining degrees or honours, is obvious. I suspect that the mathematical physicist whom I mentioned reacted to the literature which presented these kind of "results" and which, unfortunately, is not lacking in our field.

Similarly, teaching without objective aims has a tendency to become trivial and useless. If difficulties with putting across deep concepts or effective methods arise, there is no reason either to eliminate such concepts and methods from the curriculum, no matter how objectively useful they may be, or to present them in so "simplified" form that they become empty. Curiously enough, such simplifications, more often than not, are made in "service courses" under a pretext of making the subject "more suitable for applications" or "more useful". This shows to what extent the notions of usefulness and applicability have collapsed to mere convention.

There is a considerable body of scientists, including mathematicians, whose education comprised courses dealing with integration which completely disregarded the basic aims of integration theory. It would be interesting to know what percentage of instructors in basic integral calculus courses or authors of books are aware that modern integration (that is, integration since Cauchy), was invented for and tested on trigonometric series. Or how many of them are aware of the implications of their choice of integration theory (say Riemann or Lebesgue) in areas such as potential theory, stochastic processes, quantum mechanics, continuum mechanics, some branches of communication engineering, statistical mechanics, statistics, and so on. Of course, if we had no applications in mind, it would not matter what we teach. But, because the main aims for which integral calculus is being taught are outside integral calculus (nobody in his right mind would pretend to do original research in basic integration), we should teach the Lebesgue-Stieltjes integral or equivalent. But this is not usually done. What chance now has the vector measure theory even if it were the philosopher's stone curing all the ills of suffering Mankind?

I would like to finish these general remarks with one which is almost serious. There is a considerable lack of, even horror of, vector measure "spirit" in our culture (the general way of thinking). The discussion on any subject usually reduces to a quarrel about which of two extreme positions to choose or, perhaps, about which point to choose on the straight-line between the two extremes. As examples may serve: right versus left in political theory, progressive versus reactionary attitudes in ideology, rigorist versus liberal attitudes in education, centralist or parochial attitudes in organization, permissive or prohibitive morals, and so on. The discussion, whether children in schools should be made to sit straight in rows and periodically rapped on their knuckles or encouraged to jump over the heads of teaching and ancillary staff, leads nowhere as does the suggestion that we should do something in between (say, alternate the rapping of knuckles with letting them throw ink-wells on teacher). The choice is, apparently, somewhere else, and is more difficult: we have to decide when to do what. This is a serious vectorial choice.

Even when a vector is already given, it is often thrown away to be replaced by a scalar. I know of matriculation and other university examinations which result in candidates being allocated a single mark each and then linearly ordered "according to merit", although they were examined (or, rather, they sat) in several subjects, the set of subjects varying with the candidate. The procedure presupposes that excellent knowledge of Greek grammar compensates for the deficiencies in creative pottery. The examiners find it difficult to allocate to every student, as the measure of her achievement, simply the vector of marks indexed by subjects. Perhaps, not having had a thorough course in vector measures, such a solution does not occur to these worthy sages.

2. Established themes

1. Summability of functional sequences. One of the first results concerning infinite-dimensional vector measures is the discovery, in the 20's, of the phenomenon of unconditional summability of sequences in Banach spaces as distinct from the absolute summability. Undoubtedly, the most important achievement in this area is the Orlicz-Pettis lemma.

As is apparent even at this Conference, the area is not yet exhausted; new, more powerful and more general results are still being found.

The whole business started by examining the modes of convergence of orthogonal, in particular trigonometric, series. So, we have another instance of the ubiquituous stimulus derived from trigonometric series. It would, therefore, be interesting to see whether the new achievements in the direction indicated by the Orlicz-Pettis lemma could give interesting results if applied to classical problems of orthogonal series or could suggest new ones.

2. Spectral theory. If the spectral theory is the core and the summit of functional analysis, it is only fitting that it is also the main source of vector measure theory proper. The spectral theorem for a class of operators asserts that every operator in the class, or an associated operator, can be written as the integral of a scalar function with respect to a certain operator-valued measure, the spectral measure of the operator. Curiously enough, formulation in terms of spectral measures are of relatively recent origin. In the past, such theorems were formulated in terms of spectral families or similar objects analogous to distribution functions. The situation is similar to one in probability theory where measures also replaced distribution functions only relatively recently. This happened under the pressure of problems in higher dimensions where the formulation in terms of distribution functions is surely not transparent, especially when domains other than intervals are involved. Once the terminology and techniques of distribution fucntions were abandoned in favour of measure-theoretical terminology and techniques, whether in spectral or probability theory, the subject experienced quite a vigorous growth.

A superficial reason for the long survival of spectral families is that, for a single selfadjoint, normal, etc. operator, many *ad hoc* methods work and so vector-valued measures can be avoided.

On the much deeper level, there is a far-reaching analogy between positive operators in a Hilbert space and positive measurable functions with respect to a given positive measure. This analogy and its important consequences are discussed by I. Segal in the survey [45].

The conscious use of vector measures by N. Dunford lead to a considerable extension of applicability of spectral theory and to the introduction of the concept of a spectral operator in Banach or, more generally, locally convex spaces. The resulting theory was presented, in an admirable state of maturity, in the fourth volume (1954) of the Pacific Journal of Mathematics. This volume is a gold mine for a vector measure theorist even now when the Dunford's spectral theory is systematically exposed, along with subsequent developments, in the third volume of the well-known treatise [12] of N. Dunford and J.T. Schwartz.

Besides Dunford's, there are other papers in that volume of Pacific Journal showing clearly that problems of mathematical physics motivated the creation of the theory of spectral operators. In Schwartz's paper it is shown that many non-selfadjoint boundary-value problems give rise to spectral operators in the Dunford sense. W.G. Bade extends Dunford's theory to unbounded operators.

The other side of the coin is that the theory of spectral operators opened new problems for vector measure theory. S. Kakutani gives, in the same volume, a counter-example to a natural conjecture about spectral operators. His paper could be considered the first step towards the theory of tensor products of vector measures (Fubini theorem, etc.). The research in this direction does not seem to be completed. Incidentally, it only recently occurred to me, in connection with problems of perturbation of semigroups of operators (see fourth example in the third part below), that similar difficulties could be connected with the Fubini theorem for vector measures (even if one of them is scalar-valued) as with the Radon-Nikodym theorem.

The theory of spectral operators inspired one of the most important contributions to the theory of vector measures, namely that of R.G. Bartle, N. Dunford and J.T. Schwartz [3]. In my opinion, even if due is payed to the import of Grothendieck's work, the Bartle-Dunford-Schwartz result would not have come, or would not have come as early as it did, if the theory of spectral operators did not prompt it. But these matters I should, perhaps, leave to my friends [10] who discuss them thoroughly.

Spectral theory also suggests meaningful generalizations of vector measures. Kakutani's and similar examples show that the sum and product of commuting scalar-type operators are not always spectral operators. C. Foiaş used vector distributions to introduce the notions of spectral distribution and that of generalized spectral operators which do not suffer this defect. Of course, the significance of these concepts is much wider. Indeed, his work opened a new direction in spectral theory. A good reference is still the monograph [9]. Nevertheless, prominent functional analysts, including C. Foiaş (my apologies if my recollection is wrong), commented that the spectral theory ran out of fertile ideas. I believe that nonconventional applications of vector measures and their generalizations could revive this centrally important subject.

3. Moment problems. A majority of moment problems, including problems concerning integral transforms, fit in the following scheme.

Given are: a space Ω with a σ-algebra S of measurable sets in it; a set Γ, usually equipped with some structure, at least topology; a real- or complex-valued function K on $\Gamma \times \Omega$; a locally convex space X; a function $\varphi : \Gamma \to X$. The problem is, whether there exists a vector measure $m : S \to X$ such that

$$\varphi(\gamma) = \int_{\Omega} K(\gamma,\omega)dm(\omega), \qquad \gamma \in \Gamma,$$

and, possibly, how to construct it. Moreoever, we may require that m have some additional property, like bounded variation, absolute continuity with respect to a given measure, etc.

The choice $\Omega = [0,1]$, $\Gamma = \{0,1,2,\ldots\}$, $K(n,t) = t^n$, $t \in \Omega$, $n \in \Gamma$, gives the vector version of the Hausdorff moment problem. It was dealt with extensively by D. Leviatan, see e.g. [31]. Final results, based on the full strength of the Bartle-Dunford-Schwartz theorem in locally convex spaces, are due to A.K. Whitford [48].

Related to the Hausdorff moment problem is one about the Laplace transform obtained for $\Omega = [0,\infty)$, $\Gamma = [0,\infty)$, $K(\gamma,t) = \exp(-\gamma t)$, $\gamma \in \Gamma$, $t \in \Omega$. The bulk of the work was done by S. Zaidman [52]. It was finished, at least in some directions, by Whitford [47].

The case $\Omega = [a,b]$, $a < b$, $\Gamma = \{0,1,2,\ldots\}$, $K(n,x) = \varphi_n(x)$, $n \in \Gamma$, $x \in \Omega$, where $\{\varphi_n\}_{n=0}^{\infty}$ is an orthogonal system of functions on $[a,b]$, was treated by S.K. McKee [34]. An interesting instance is obtained if φ_n, $n = 0,1,2,\ldots$. are the eigenfunctions of a Sturm-Lioville operator. He treated, in [35], also the continuous analogue of this case, i.e. one of an unbounded interval instead of $[a,b]$. Considering the potential use in boundary and initial value problems of mathematical physics, McKee's results, I feel, deserve an investment of further work.

The case $\Omega = \mathbb{R}^l$, $\Gamma = \mathbb{R}^l$, $K(\gamma,x) = \exp(i\gamma x)$, $\gamma \in \mathbb{R}^l$, $x \in \mathbb{R}^l$, should serve as a paradigm for any pair of locally compact Abelian groups in duality. I know of two ways to solve the corresponding (trigonometric) moment problem.

One is a vector extension of the Bochner-Schoenberg-Eberlein test. Namely, a weakly continuous function $\varphi : \mathbb{R}^l \to X$ can be expressed in the form

$$(1) \qquad \varphi(\gamma) = \int_{\mathbb{R}^l} \exp(i\gamma x)dm(x), \qquad \gamma \in \mathbb{R}^l,$$

with an X-valued measure m on $\mathbb{R}^l$ if and only if the set

$$\{ \int_{\mathbb{R}^l} \varphi(\gamma)f(\gamma)d\gamma \; : \; f \in L^1(\mathbb{R}^l), \; \|\hat{f}\|_\infty \le 1\}$$

is relatively weakly compact in X; $\hat{f}$ denotes the Fourier transform of f. The absolute continuous measures $f(\gamma)d\gamma$ can be replaced in this criterion by ones with finite supports. This was stated in [21] for the Banach space X. As several authors noted, the general situation is analogous.

The other way uses some positive kernel (approximate identity) such as Fejér's or Riesz's. Let

(2) $$p(x,t) = \frac{1}{(2\pi t)^{\frac{1}{2}l}} \exp\left(-\frac{|x|^2}{2t}\right), \quad x \in \mathbb{R}^l, \ t > 0.$$

Let $\varphi : \mathbb{R}^l \to X$ be a bounded weakly continous function. Let

$$F_t(x) = \int_{\mathbb{R}^l} \exp(i\gamma x) p(\gamma,t) \varphi(\gamma) d\gamma, \quad x \in \mathbb{R}^l, \quad t > 0.$$

There exists an X-valued measure m on $\mathbb{R}^l$ such that (1) holds if and only if, for every $t > 0$ the function F_t is integrable (with respect to the Haar=Lebesgue measure), and the maps $\Phi_t : C_0(\mathbb{R}^l) \to X$, $t > 0$, defined by

$$\Phi_t(f) = \int_{\mathbb{R}^l} f(x) F_t(x) dx, \quad f \in C_0(\mathbb{R}^l), \quad t > 0,$$

are weakly compact uniformly with respect to $t > 0$ (a weakly compact set exists, not depending on t, into which the unit ball of $C_0(\mathbb{R}^l)$ is mapped by every Φ_t, $t > 0$).

This method, using approximate identity, can be used also for finding out whether m has finite variation, or whether it has Radon-Nikodym derivative with respect to the Haar measure, [22].

Besides my obvious bias, the reasons for mentioning the solutions of the trigonometric moment problem is that it has several direct uses. One which immediately comes to mind is a generalization of Stone's theorem on spectral representation of unitary groups. Any unitary group of operators $t \mapsto U_t$, $t \in \mathbb{R}^1$, can be written in the form

(3) $$U_t = \int_{-\infty}^{\infty} \exp(itx) dM(x), \quad t \in \mathbb{R}^1,$$

where M is a spectral measure (of course, the parameter-group can be much more general than $\mathbb{R}^1$). The operator-valued measure M is the solution of a trigonometric moment problem. Hence, we can consider other groups of operators, not only unitary ones. This is, of course, a possible approach to spectral theory, that is, for finding sufficient conditions for spectrality of an operator [21].

Similarly, if $t \mapsto U_t$, $t \in \mathbb{R}^1$, is a stationary second order process, then there exists an orthogonally scattered measure M such that (3) holds.

Its values are interpreted, of course, as random variables. To M. Loève [33] is attributed the introduction of a wider class of processes, called harmonizable processes, having in many aspects similar properties to the stationary ones. They were studied by several authors, notably by H. Cramér. Vector measures were introduced into this business by Yu. A. Rozanov who, for this purpose, developed independently some portions of vector measure theory. In a recent work [41], A.G. Miamee and H. Salehi use the full strength of vector measure theory to obtain a nice representation of harmonizable second order processes. I refer to their paper for the details and for basic references which I omitted (including ones to Cramér, Rozanov, Masani, Naimi and others).

When writing this, it occurred to me that it would be interesting to examine the relation between the Miamee-Salehi results and Ambrose's generalization of Stone's theorem [1]. More generaly, the mutual implications of spectral theories of operators and stochastic processes may not yet have been sufficiently explored.

4. Conditional probabilities. Let (Ω, S, P) be a probability space. Let T be a sub-σ-algebra of S. The conditional probability with respect to T is a function $P_T = P_T(\cdot|\cdot)$ on $S \times \Omega$ such that

(i) for every $E \in S$, the function $\omega \mapsto P_T(E|\omega)$, $\omega \in \Omega$, is T-measurable, and

(ii) for every $E \in S$ and $F \in T$,

$$P(E \cap F) = \int_F P_T(E|\omega) dP(\omega).$$

It follows that $E \mapsto P_T(E|\cdot)$, $E \in S$, is a map from S into T-measurable functions (random variables, if you prefer) which is σ-additive in the sense of almost sure convergence. This is to say, if E_n, $n = 1,2,\ldots$, are pairwise disjoint sets from S with union E, then the equality

$$P_T(E|\omega) = \sum_{n=1}^{\infty} P_T(E_n|\omega) \tag{4}$$

holds for P-almost every $\omega \in \Omega$. However, it is not possible, in general, to find a set $N \in S$ such that $P(N) = 0$ and (4) holds for every $\omega \notin N$, every $E \in S$ and every partition $\{E_n\}_{n=1}^{\infty}$ of E into pair-wise disjoint sets from S. Hence, whether we want it or not, $E \mapsto P_T(E|\cdot)$, $E \in S$, must be treated as a vector measure whose values are T-measurable functions and not as a family of ordinary probability measures indexed by almost all elements of Ω. Because the values of this vector measure are uniformly bounded, we can treat the measure as L^p-valued, for any $p < \infty$. For $p = \infty$, the σ-additivity cannot be in norm. Because the space L^0 (the space of all measurable functions equipped with metric inducing the convergence almost everywhere) is not locally convex, we have one of the relatively rare instances when the attention payed to non-locally-convex space valued measures may be warranted.

A considerable amount of effort has been spent on determining whether a given conditional probability can be "unvectorized". We seek conditions guaranteeing that, for P-almost $\omega \in \Omega$, the function $E \mapsto P_T(E|\omega)$, $E \in S$, is a (numerical-valued) measure. The first result in this direction, as far as I know, is due to M. Jiřina [17]. The problem is often treated by methods of lifting. Chatterji's contribution [8] to Snowbird Symposium is a good reference.

I can't resist making a small remark here. Professor McShane reports in his survey [36], p. 599, a nice theorem, due to Shu-Teh-Chen Moy and Gian-Carlo Rota, stating that every averaging operator on $L^p(\Omega,S,P)$, $p \in [1,\infty]$, is necessarily a conditional probability operator. He remarks that averaging operators do not at a glance look much like integrals. He goes on further saying that he does not "feel completely certain as to whether or not a theorem about averaging operators can justly be called a theorem about integration." I wish to suggest that theorems about averaging operators are indeed theorems about integration. The only point is that it is a vector measure with respect to which one integrates, namely the conditional probability vector measure which I just described. Surely, Professor McShane is completely at ease as to whether the Moy-Rota theorem is interesting and important or not and justly considers a question of its classification as secondary, but he gives me an opportunity to put in my usual commercial slogan: "Vector measures to every household."

5. Stochastic integration. Let $t \mapsto W_t$, $t \geq 0$, be the Wiener process with variance σ^2 per unit time. Let it be realized on the probability space (Ω, S, P); it is irrelevant, for our discussion, that the set of all sample paths can be taken for Ω. Let f be a real-valued function on $[0,1]$, say. The difficulties with the definition of the integral

$$\int_0^1 f(t)dW_t \tag{5}$$

are due to the fact that, for P-almost every $\omega \in \Omega$, the function $t \mapsto W_t(\omega)$ has infinite variation in every interval $[\alpha,\beta]$, $0 \leq \alpha < \beta$. However, there exists a unique $L^2(\Omega,S,P)$-valued measure m on the δ-ring of Borel sets in $[0,\infty)$ of finite Lebesgue measure such that

$$m((\alpha,\beta]) = W_\beta - W_\alpha,$$

for any $0 \leq \alpha < \beta$. For every set E in the domain of m, the value $m(E)$ is a Gaussian random variable with zero expectation and with variance, $\|m(E)\|^2$, equal to σ^2 times the Lebesgue measure of E. On disjoint sets the value of m are independent and, hence, orthogonal. So the measure m is orthogonally scattered.

The integral (5) can now be defined simply as the integral of the function f with respect to the vector measure m. For use in applications, the vector measure m may actually seem a more natural object than the Wiener process itself. This is more apparent in dimensions higher than 1. The analogue of the measure m in higher dimensions occurred already in the 30's in Wiener's work [49] on homogeneous chaos. For a probabilistic model of an instantaneous state of homogeneous turbulence in $\mathbb{R}^3$, it is natural to take an orthogonally scattered $L^2(\Omega,S,P)$-valued measure m defined on all Borel sets of finite Lebesgue measure such that the values of m are Gaussian random variables with zero expectation and variance proportional to the Lebesgue measure.

We see that, while it may be difficult to extend satisfactorily the Markov property to processes with several-dimensional time parameter (random fields), there are no substantial difficulties with defining random fields "with independent increments". Again, problems in higher dimensions force the measure-theoretic methods into the fore.

There is a considerable literature devoted to problems of stochastic integration. The vector-measure-theoretic viewpoint is taken openly by M. Métivier, J. Pellaumail and others. Relatively recent progress in this direction is reported in the monograph [29]. The subject is far from exhausted. It poses still new non-trivial problems as can be seen in recent works of Métivier and Pellaumail [40] or M. Yor [51].

As all working probabilists would confirm, the real problems, from their point of view, start when also the values of the function f depend on chance. This situation, though of utmost importance, is far from satisfactorily dealt with. This is not to deny the successes of procedures like the Itô integral prompted by and satisfactorily solving various problems concerning stochastic differential equations. These procedures obtained already some thorough treatment in the literature, for example in the well-known book [37] of McShane. But still, it has to be admitted that many methods have an *ad hoc* character; they depend too much on specific properties of processes involved, for example on the martingale property. Consequently, these methods are hard to generalize satisfactorily to dimensions higher than 1.

The vector measure approach may contribute here to some progress. The integral (5), in the case that both functions f and $t \mapsto W_t$ are random, should be treated as a bilinear integral, both the integrand and the measure are vector-valued.

Now, the first problem we are facing in this situation is the choice of the bilinear operation (tensor product) pairing the values of the function and the measure. This is not a matter of indifference. A judicious choice of the tensor product can improve theoretical properties of the integral and increase considerably the possibility of success in applications. This was nicely shown by Steel Huang in his recent colloquium talk in the Department of Statistics. I understand (and hope) that his work is to be published soon.

The second problem is that the theory of bilinear integral has not progressed sufficiently since Bob Bartle's paper [3]. I. Dobrakov [11] contributed to our understanding of the issue but I feel the situation is still not sufficiently understood. In the case of scalar functions and scalar or vector measures we can, clearly, interpret the integration theory as concrete representation of completions of some function spaces (of continuous functions or step-functions) with respect to convergence in the mean. The completeness of L^p and the density of some sets of functions in it are responsible for the main successes of Lebesgue integration theory and for the possibility of formulating it both concisely and independently of any particular choice of integration procedure. I feel that until a similar state of perfection is achieved, it will be difficult to use bilinear integrals with ease and, at the same time, effectively.

I wish to add, on an optimistic note, that bilinear integration has already scored considerable successes in the fields of application; I will mention some in the point 8 below. Secondly, the problems related to it attract attention and we may expect considerable progress soon. For example, Professor Kakutani informed me, after my talk, that he is working with a student on problems related to these themes.

6. The range. Nontrivial problems concerning vector measures with bounded variation do exist. The Radon-Nikodym theorem is an example. Questions related to the range go further. Here we find nontrivial problems even for finite-dimensional vector measures. Indeed, they give rise to "the most beloved theorem on vector measures", namely the Liapunov theorem stating that the range of a non-atomic finite-dimensional vector measure on a σ-algebra is convex and compact.

Already the two-dimensional version is interesting. It is the core of the Neyman-Pearson lemma, dated from the early 30's. The motivation for Liapunov still derives from statistics. At least this impression is obtain from reading the original paper [32]. It may be interesting to note that Liapunov was advised by A.N. Kolmogorov. The usage in statistical theory is explained, in modern terms, for example, in Lehman's monograph [30].

After the war the interest of statistician still persisted [13]. The most significant contributions for the extending of this theme beyond statistics seem to be that of Blackwell [7] and Karlin [19]. Karlin introduces integration of vector functions with respect to vector measures into the play (still in finite dimensions). He seems to be motivated by optimization problems; it was long before these problems became fashionable. His influence goes very far; his ideas are behind the well-known Lindenstrauss proof of Liapunov's theorem and also behind many recent works on optimization. The literature on the subject is vast.

A new surge of interest in the Liapunov theorem was caused by the appearance of the work of J.P. LaSalle who has put the "bang-bang principle", used by engineers intuitively for some time, on a rigorous basis. The basic idea behind it can be simply derived from Liapunov's theorem. This contributed substantially to the understanding of the bang-bang principle and caused, in fact, a deluge of works on the theme. Not all of them were of the highest merit, many were devoted simply to new proofs of Liapunov's theorem or its variants, but there occurred also many penetrating works contributing to the clarification of principles involved and also to the unification of approaches to optimization problems of engineering and economic origins.

It was noted already by Liapunov himself that his theorem, as it stands, fails in infinite-dimensional spaces. The research in this direction was desultory until the occurrence of the paper [20] by J.F.C. Kingman and A.P. Robertson. It was rounded off by the formulation of a general version of the Liapunov theorem due to Greg Knowles. I refer to Chapter V of [27] for a precise formulation and comments. The survey [28] indicates the ways of using the general Liapunov theorem in control of systems with distributed parameters.

I wish to suggest also that an infinite-dimensional Liapunov theorem existed, long before its precise formulation, implicitly in engineering literature. The St. Venant's principle, used in elasticity since the last Century, is more-or-less an assumption that for certain vector measures (of course, not identified explicitly) the statement of the Liapunov theorem holds. A few further remarks, perhaps more illuminating, can be found in Section V.7 of [27].

The proof that a given vector measure is Liapunov (the restriction to every measurable set has weakly compact and convex range) can often be based on the lacunarity of some functional series. It would be delightful to obtain results on lacunary series using vector measures, specifically Greg Knowles's theorem. At any rate, there is a strong interplay which brings forward some nice and, it seems, hard problems with far reaching implications. Here is one.

Let $\{n_k\}_{k=1}^{\infty}$ be an increasing sequence of positive integers such that $\sum n_k^{-1} < \infty$. Decide whether, for every Borel set $E \subset [0,1]$ with nonzero Lebesgue measure, there exists a bounded measurable function f such that

$$\int_E |f(t)|\,dt \neq 0, \text{ and } \int_E f(t)t^{n_k}dt = 0,\ k = 1,2,\ldots.$$

If a vector measure fails to be Liapunov, what then? If we restrict attention to a Banach space valued measure m on a σ-algebra S of sets in a space Ω, then we can guarantee the existence of pair-wise disjoint measurable sets $E_0, E_1, E_2, \ldots$ with union Ω such that, the restriction of m to no non-negligible subset of E_0 is Liapunov and the restriction of m to each $E_1, E_2, \ldots$ is Liapunov.

This kind of result could be prompted simply by a desire to tidy things up. But, as a matter of fact, "these results find applications in stochastic models of economic equilibrium" [14].

After considering the Liapunov theorem, the question naturally arises, which weakly compact convex sets are the range of a vector measure? In finite-dimensional spaces, the question was answered by N.W. Rickert [42] by showing that they are precisely zonoids containing the zero-vector of the space. This characterization was extended to infinite-dimensional spaces in [24]; it is reported in Chapter VII of [27].

The whole problem is rather attractive because it ties together ideas from various disciplines: zonoids from geometry, Lévy's stable laws from probability, negative-definite functions from harmonic analysis, and, of course, vector measures. Control theory was thrown into the bargain by H.S. Witsenhausen by his remark [50]. The results about the range of vector measures are results about the attainable set of linear control systems. One, perhaps surprising, consequence of their characterization as zonoids is that the attainable set of a system with many controls operating independently is the attainable set of another system with a single control (of course, the new system could be rather wild).

7. Representation of mappings. Natural applications of vector measures, once we have them, are representations of maps from various function spaces into a locally convex space X as integration with respect to X-valued measures. This includes various generalizations of the Riesz representation theorem of functionals on continuous function spaces and representations of maps from L^p spaces into X. These questions though very interesting, are not directly related to the theme of this essay, therefore I confine myself to the recommendation of two detailed surveys [5] and [6] by Jürgen Batt.

He mentions in these surveys also some cases of representation of nonlinear maps and gives corresponding references. Such cases may be closer to the type of applications we have in mind here.

Recent developments brought very interesting and, to me, surprising results. The surprising element is their wide applicability. By a simple algebraic rearrangement, almost like a conjurer's trick, almost any operator of interest reduces to a disjointly additive operator. Now, disjointly additive operators are effectively represented using vector measures. The vector-measure-theoretician's delight is the inherent use of Liapunov's theorem and other methods close to our heart. I refer to papers [38] and [39] by M. Marcus and V.J. Mizel for details and further references. Anyway, they contributed the lion's share to this recent developments. It may be of interest to note the influence of Blackwell's ideas in their work. It seems a pity that his paper is not better known.

8. Bilinear integration. I suggested that the bilinear integration is not yet in a perfect state of completion. Nevertheless, it has already provided useful services.

It was interesting to listen to the late Professor Jauch, at the Snowbird Symposium describing his efforts and those of his collaborators to clarify the exact relationship between the stationary and the time-dependent scattering theories of quantum mechanics. For this purpose he needed to integrate vector-valued functions with respect to vector-valued measures. Hence, he applied himself to the task of developing a corresponding theory. However, in the mean-time he discovered that there already existed one, namely that of R.G. Bartle [3]. So he used it with a success proving ultimately the equivalence of the two scattering theories. Some people tend to attribute to this achievement similar importance as to showing the equivalence of Heisenberg's matrix mechanics with Schrödinger's wave mechanics. At any rate, it is pleasing to read the paper [2], if for nothing else, than to see the advertisement of vector-measure-theoretical methods of prominent workers in a field seemingly so remote from vector measures and so important as quantum mechanics.

A parallelism between quantum mechanics, in particular scattering theory, and some aspects of communication theory was already noted. I do not know, besides some very technical notes on the subject, about an accessible reference. Moreover, I do not understand the exact nature of this analogy. But it may be significant that in some problems of realizability theory bilinear integration enters in a similar fashion as in scattering theory. For this, I would like to refer to the monograph [53] of A.H. Zemanian. Maybe a progress in bilinear integration will contribute to improvement of methods to be used in quantum mechanics and communication theory on one side and may lead to a complete clarification of their mutual relation on the other.

3. Examples

1. Fourier-Stieltjes transforms. Let C be the Banach space of bounded continuous functions on $\mathbb{R}^1$. Given $f \in C$ and $\gamma \in \mathbb{R}^1$, let $f_\gamma(\xi) = f(\xi+\gamma)$, for $\xi \in \mathbb{R}^1$. Using the first solution of the trigonometric moment problem (in the form with discrete measures) mentioned in point 3 of part 2, we see that there exists a C-valued measure m on $\mathbb{R}^1$ such that

$$f_\gamma = \int_{\mathbb{R}^1} \exp(i\gamma x)dm(x), \quad \gamma \in \mathbb{R}^1, \tag{6}$$

if and only if

$$\left\{ \sum_{j=1}^{m} c_j f_{\gamma_j} : \sup\left\{ \left| \sum_{j=1}^{n} c_j \exp(i\gamma_j x) \right| : x \in \mathbb{R}^1 \right\} \leq 1 \right\} \tag{7}$$

is a relatively weakly compact subset of C.

Now, (6) holds for some C-valued measure m on $\mathbb{R}^1$ if and only if, there exists a complex-valued measure μ on $\mathbb{R}^1$ such that

$$f(\gamma) = \int_{\mathbb{R}^1} \exp(i\gamma x)d\mu(x), \quad \gamma \in \mathbb{R}^1. \tag{8}$$

Indeed, if μ is given, satisfying (8), we define m by

$$m(E)(\gamma) = \int_{\mathbb{R}^1} \exp(i\gamma x)d\mu(x), \quad \gamma \in \mathbb{R}^1,$$

for every Borel set E, and m will satisfy (6). If m is given, we define μ by $\mu(E) = m(E)(0)$, for every Borel E, and μ will satisfy (8).

So we proved that *a necessary and sufficient condition for the existence of a complex measure* μ *on* $\mathbb{R}^1$ *such that* (8) *holds is the weak compactness of the set* (7).

A similar argument applies in the case of any pair of locally compact Abelian groups in duality instead of $\mathbb{R}^1$ and $\mathbb{R}^1$ [23]. I included this simple argument to give an example of a proof of a statement, which has nothing to do with vector measures using vector-measure-theoretical method. The statement can, of course, be proven without involving vector measures. But, I must confess, it would not have occurred to me without thinking about vector measures (one direction was of course known). We can make analogous statement using absolutely continuous measures, that is L^1, instead of discrete ones. The result inspired some modest follow-up. A similar treatment given to other integral transforms could, perhaps, produce some interesting results.

2. Cylindrical measures. Let X be a quasi-complete locally convex space. Let $\mathcal{B}$ be the algebra of sets of the form $\varphi^{-1}(E)$, for every continuous linear map $\varphi : X \to \mathbb{R}^k$, every Borel set $E \subset \mathbb{R}^k$, and $k = 1,2,\ldots$. A function μ on $\mathcal{B}$ is called a cylindrical measure if, for any given continuous linear map $\varphi : X \to \mathbb{R}^k$ the restriction of μ to the σ-algebra of all sets $\varphi^{-1}(E)$ with Borel $E \subset \mathbb{R}^k$ is a σ-additive probability measure. The main question concerning a cylindrical measure is its σ-additivity.

Let (Ω, S, P) be a probability space. Let $m : S \to X$ be a vector measure such that $m \ll P$. Then the cylindrical measure μ defined on $\mathcal{B}$ by

$$\mu(\varphi^{-1}(E)) = P((d\varphi \circ m/dP)^{-1}(E)),$$

for any continuous linear map $\varphi : X \to \mathbb{R}^k$, any Borel set $E \subset \mathbb{R}^k$, and any $k = 1,2,\ldots$, is called the (m,P)-distribution on X.

The (m,P)-distribution has an obvious interpretation. We pretend that the Radon-Nikodym derivative $f = dm/dP : \Omega \to X$ exists and then take the distribution of f interpreted as an X-valued random variable. Exactly because the existence of the RN derivative is only a pretence, we use the cautious term "(m,P)-distribution" instead of the "distribution of dm/dP."

We say that the vector measure m (we still assume $m \ll P$) is virtually P-differentiable in X if there exists a σ-algebra $\tilde{S} \supset S$ and a probability measure $\tilde{P}$ on $\tilde{S}$ such that

(i) $\tilde{P}(E) = P(E)$, for every $E \in S$;

(ii) for every $E \in \tilde{S}$, there is $F \in S$ such that $\tilde{P}(E \triangle F) = 0$;

(iii) the unique vector measure $\tilde{m} : \tilde{S} \to X$, extending m, such that $\tilde{m} \ll \tilde{P}$ has a Pettis integrable density $\tilde{f}$ with respect to $\tilde{P}$ with values in X.

This notion finds its use in the following statement.

If the vector measure m is virtually P-differentiable, then the (m,P)-distribution is σ-additive.

Indeed, let μ be the (m,P)-distribution. Then $\mu(\varphi^{-1}(E)) = \tilde{P}(\tilde{f}^{-1}(\varphi^{-1}(E)))$, for every continuous linear map $\varphi : X \to \mathbb{R}^k$, every Borel set $E \subset \mathbb{R}^k$, and $k = 1,2,\ldots.$ The σ-additivity of μ then follows.

This simple statement is the basis of a scheme for using Radon-Nikodym theory in probability, namely in construction of generalized stochastic processes. The classical Doob construction of the Wiener process fits into it [25]. The statement has also a converse which, however, must be more carefully formulated than in the note [25]. The error was removed by Brian Jefferies.

When I wrote the note [25], I too wallowed in ignorance. RN theory was already used for these problems by L. Schwartz [44] and, more recently, by A. Goldman [15].

3. Boundary value problems. As an illustration, let us observe first how the solutions of the problem

(9) $$\frac{\partial u}{\partial t} = \frac{\partial^2 u}{\partial x^2},\ (x,t) \in (0,\pi)\times(0,\infty);\ u(0,t) = u(\pi,t) = 0,\ t \in (0,\infty);$$

depends on the initial-value condition

(10) $$u(x,0) = f(x), \quad x \in [0,\pi].$$

Let X be the space of all continuous functions $(x,t) \mapsto u(x,t)$, $(x,t) \in [0,\pi]\times(0,\infty)$, vanishing at $x = 0$ and $x = \pi$, with continuous derivative with respect to t and continuous first and second partial derivatives with respect to x, equipped with the topology of uniform convergence of functions and partial dervatives involved on every interval $[0,\pi]\times[\alpha,\infty)$, with $\alpha > 0$.

If f is a continuous function in $[0,\pi]$ such that $f(0) = f(\pi) = 0$, then the problem (9) and (10) has a unique solution u continuous in the closed interval $[0,\pi]\times[0,\infty)$. Let $m(f)$ be the restriction of this solution to the interval $[0,\pi]\times(0,\infty)$ interpreted as an element of the space X. The maximum principle implies that this defines an X-valued Daniell integral or, if you prefer, an X-valued Radon measure m on the space of all continuous functions in $[0,\pi]$ vanishing at 0 and at π. The vector measure m solves the problem (9) for various initial conditions (10); the solution of (9) and (10) is given as the integral of f with respect to m. In this way, many generalized solutions, corresponding to non-continuous initial-value conditions f, can be obtained. If the function f in (10) is m-integrable, then the problem (9), (10) surely admits a generalized solution in a good sense. This idea, though classical, is actually more acceptable if we think of the problem in its original physical terms of conduction of heat or diffusion rather than in terms of a partial differential equation.

We can go on further. The vector measure m has the Radon-Nikodym derivative $\xi \mapsto G(\xi)$, $\xi \in [0,\pi]$, with respect to Lebesgue measure, so that

$$m(f) = \int_0^\pi f(\xi)G(\xi)d\xi,$$

for every m-integrable function f on $[0,\pi]$. The derivative G can be given explicitly in this case:

$$G(\xi)(x,t) = \frac{2}{\pi}\sum_{n=1}^{\infty} \exp(-n^2t)\sin nx \sin n\xi,\ \ \xi \in [0,\pi],\ \ (x,t) \in [0,\pi]\times(0,\infty).$$

Now we can consider generalized solutions of (9) subject to initial conditions given by some measures on $[0,\pi]$ rather than functions. If the integral

$$\int_0^\pi G(\xi)d\mu(\xi)$$

exists as an element of X (in the Bochner or Pettis sense), then this element is the solution of (9) subject to the initial-value condition given by the measure μ.

The Green's function can be obtained similarly for many other boundary and/or initial value problems as the Radon-Nikodym derivative of the vector measure which solves the problem for "all" boundary or initial conditions, with respect to a "natural" measure on the boundary such as the surface area. Eric Thomas pointed in [46] at this important application of RN derivatives. To be sure, the values of the RN derivative may not belong to the space of solutions, but to some of its extensions such as the second dual, or, in bad cases, the algebraic dual of the continuous dual.

I believe that research in this direction is worth pursuing further. And I wish to make still another comment. There may be difficulties with finding the RN derivative. Also, a "natural" measure on the boundary may not be available or it may not be opportune to bring it into the play. In such cases, the vector measure itself should suffice for the analysis. The Green's function (RN derivative) only reduces the integration with respect to the vector measure to integration with respect to a positive measure deemed to be more familiar. To insist on using Green's function cost what it may, means paying undue respect to conventions and may jeopardise success in solving a problem.

A classical instance of working with a vector measure without involving the Green's function is the Perron-Wiener-Berlot method for finding generalized solutions of the Dirichlet problem; see, for instance, Chapter 8 of the monograph [16]. I do not wish to accuse Perron, Wiener or Berlot of using vector measures openly, but the method can be thought of as an instance of a vector Daniell integral, one whose values are harmonic functions. The properties of harmonic functions, especially ones involving order, may obscure the fact that what is going on is in fact some sort of an integration procedure.

It would be interesting to pursue this path in other boundary or initial value problems for which the maximum principle is not available. In such cases, the techniques made possible by the specific properties of harmonic functions will have to be replaced by the methods of vector integration.

4. Perturbation of semigroups. Let X be a locally convex space. Let $L(X)$ be the space of all continuous linear operators. Let $S : [0,\infty) \to L(X)$

be a continuous semigroup of operators. Assume that Λ is a locally compact Hausdorff space, P a spectral measure on (Baire sets of) Λ, and V a Baire function on Λ. Let

(11) $$T(t) = \exp\left(t\int_\Lambda V(\lambda)dP(\lambda)\right), \quad t \in [0,\infty).$$

The semigroup S describes an evolution process in which any element φ of the space X evolves during a time-interval of duration $t \geq 0$ into $S(t)\varphi$. Another process is described by the map $T : [0,\infty) \to L(X)$. This is to say, we assume that the operators $T(t)$ are well-defined by (11) and belong to $L(X)$, for every $t \geq 0$, and that any vector $\varphi \in X$ evolves in time $t \geq 0$ by the action of the semigroup T into the vector $T(t)\varphi$. More generally, if the operators (11) do not belong to $L(X)$, we consider the evolution only of vectors $\varphi \in X$ such that $T(t)\varphi \in X$, for every $t \geq 0$.

The problem is to determine the vector of the space X into which a given vector $\varphi \in X$ evolves in a time $t \geq 0$ if both processes go on simultaneously. In other words, we wish to construct a new process which is the superposition of the two processes. I suggest the following solution.

Given $t > 0$, let $\Gamma_t = \{\gamma : [0,t] \to \Lambda,\ \gamma \text{ continuous}\}$ be the space of all continuous paths in Λ. For a set

(12) $$E = \{\gamma : \gamma \in \Gamma_t,\ \gamma(t_j) \in B_j,\ j = 1,2,\dots,n\},$$

where $0 \leq t_1 < t_2 < \dots < t_n = t$ and $B_1, B_2, \dots, B_n$ are Baire sets in Λ, we let

$$M_t(E) = S(t-t_n)P(B_n)S(t_n-t_{n-1})P(B_{n-1})\dots P(B_2)S(t_2-t_1)P(B_1)S(t_1).$$

If M_t extends to an $L(X)$-valued measure on the σ-algebra S_t generated by all sets (12), we shall call the extended measure the (S,P,t)-measure and shall denote it still by M_t. To be sure, the (S,P,t)-measure may not exist. If it does, it means that, for every $\varphi \in X$, the application $E \mapsto M_t(E)\varphi$, $E \in S_t$, is an X-valued σ-additive measure on S_t.

Assume that, for every $t \geq 0$, the (S,P,t)-measure $M_t : S_t \to L(X)$

exists. If the integral

$$(13) \qquad U(t) = \int_{\Gamma_t} \exp\left[\int_0^t V(\gamma(r))dr\right] dM_t(\gamma)$$

exists, for every $t \geq 0$, as an element of $L(X)$, then $U : [0,\infty) \to L(X)$ is a semigroup which is the superposition of the two semigroups S and T. This is not difficult to see directly by a standard "physical" argument, namely by chopping the interval $[0,t]$ into small pieces and letting S and T act alternately. Using a vector version of the Fubini theorem [26], we can relate the action of U to more classical problems of integral or differential equations.

Given $\varphi \in X$, let us denote by the vector measure $E \mapsto M_t(E)\varphi$, $E \in S_t$, by $M_t\varphi$. Let leb denote the Lebesgue measure on any interval in $\mathbb{R}^1$; by $M_t\varphi\otimes\text{leb}$ is denoted the tensor product of measures leb and $M_t\varphi$. The symbol $S'^*(s)$ will denote the algebraic adjoint of the adjoint of the operator $S(s)$, $s \geq 0$. Finally, $P(V)$ denotes the integral of a function V on Λ with respect to the spectral measure P. Using these notations we can conveniently make the following statement.

If $\varphi \in X$ and if the function f defined by

$$f(\gamma,q) = V(\gamma(q))\exp\left[\int_0^q V(\gamma(r))dr\right], \quad \gamma \in \Gamma_t, \ q \in [0,t],$$

is $M_t\varphi\otimes$leb-integrable on $\Gamma_t\times[0,t]$, for every $t \geq 0$, then the integral

$$(14) \qquad u(t) = \int_{\Gamma_t} \exp\left[\int_0^t V(\gamma(r))dr\right] dM_t(\gamma)\varphi$$

exists as an element of X and the equality

$$(15) \qquad u(t) - \int_0^t S'^*(t-q)P(V)u(q)dq = S(t)\varphi$$

holds for every $t \geq 0$.

To obtain (15), one integrates the function f first with respect to leb and then with respect to $M_t\varphi$ and then one integrates with respect to $M_t\varphi$ first and by leb afterwards [26].

If $P(V)u(q) \in X$, for almost every $q \in [0,t]$, then the equation (15) can be written as

$$(16) \qquad u(t) - \int_0^t S(t-q)P(V)u(q)dq = S(t)\varphi .$$

If, moreover, $u(t)$ belongs to the domain of the infinitesimal generator A of the semigroups S, then $t \mapsto u(t)$, $t \geq 0$, is a solution of the initial-value problem

$$\dot{u}(t) = Au(t) + P(V)u(t), \ t > 0; \ \lim_{t\to 0+} u(t) = u(0) = \varphi .$$

The fundamental solution of this problem is the semigroup $t \mapsto U(t)$, $t \geq 0$, given by (13).

Let us consider a classical special case. We take $\Lambda = \mathbb{R}^l$, say $l = 3$. The space X consists of all σ-additive scalar-valued measures on (Borel sets in) $\mathbb{R}^l$. The semigroup $S : [0,\infty) \to L(X)$ is defined by letting

$$(S(t)\mu)(B) = \int_B dy \int_{\mathbb{R}^l} p(x-y,t)d\mu(x),$$

for every $t > 0$, every $\mu \in X$ and every Borel set $B \subset \mathbb{R}^l$; the kernel p is defined by (2).

For every Borel $B \subset \mathbb{R}^l$ and $\mu \in X$, the symbol $P(B)\mu$ will represent the measure such that $(P(B)\mu)(C) = \mu(B\cap C)$, for every Borel set $C \subset \mathbb{R}^l$. This defines a spectral measure P on $\mathbb{R}^l$.

For every $t \geq 0$, the (S,P,t)-measure $M_t : S_t \to L(X)$ does exist. This can be quite easily deduced from the existence of the Wiener measure. The (S,P,t)-measures are even σ-additive in the operator-norm topology.

If μ is a probability measure, the measure $S(t)\mu$ is a probability measure again, representing the probability distribution of the position at time $t \geq 0$ of a Brownian particle whose position at time 0 was distributed according to the law μ. Alternatively, if μ represents the mass distribution of a substance in the space filled by some solvent, then $S(t)\mu$ represents the distribution of the substance after it was left to diffuse spontaneously for a time $t \geq 0$.

Assume that the substance is also created (say, by a reaction within the solvent) at a rate proportional to the amount already present, the coefficient of proportionality $V(x)$ depending on the place $x \in \mathbb{R}^l$ in the space. If $V(x) \leq 0$ one should more properly speak about the destruction of the substance. It is easy to imagine such situations by considering some radioactive substances, say.

In probabilistic terms, we assume that the Brownian particle is exposed to a risk of destruction characterized by a non-positive function V on $\mathbb{R}^l$. The meaning of V is given by saying that, if the particle is at the point x then the probability that it will survive there for a short period of time, τ, is equal to $1 + \tau V(x) + o(\tau)$, $\tau \to 0$.

It is clear that, if the function V has non-positive values, it should be possible to find the distribution of the substance at any time $t \geq 0$, if the diffusion and the destruction processes are going on simultaneously. Or, for any $t \geq 0$ and any Borel set $B \subset \mathbb{R}^l$, there must be a number $\nu_t(B)$ representing the amount of the substance in the set B at time t. This number may also mean the probability of finding the Brownian particle in the set B at the time t if it started its Brownian journey with distribution μ in an environment with destruction risk characterized by the function V. Indeed, the formula

(17) $$\nu_t = \int_{\Gamma_t} \exp\left(\int_0^t V(\gamma(r))dr\right) dM_t(\gamma)\mu, \quad t \geq 0,$$

gives the answer. The measures ν_t can be calculated from the integral equation

$$\nu_t - \int_0^t S'^*(t-q)V\nu_q dq = S(t)\mu, \ t \geq 0.$$

If $V\nu_q$ belongs to X, for almost every $q \in [0,t]$, we can write

(18) $$\nu_t - \int_0^t S(t-q)V\nu_q dq = S(t)\mu, \quad t \geq 0.$$

The measures ν_t have densities $x \mapsto u(x,t)$, $x \in \mathbb{R}^l$, with respect to the Lebesgue measure, for every $t > 0$. Hence the equation (18) can be written as

$$u(x,t) - \int_0^t \int_{\mathbb{R}^l} p(x-y,t-q)V(y)u(y,q)dydq = \int_{\mathbb{R}^l} p(x,-y,t)d\mu(y)$$

for any $x \in \mathbb{R}^l$ and $t > 0$. This, in turn, is an integral form of the problem

(19) $$\frac{\partial u}{\partial t} = \tfrac{1}{2}\Delta u + Vu, \ t > 0, \ x \in \mathbb{R}^l; \ \lim_{t\to 0} \int_B u(x,t)dx = \mu(B),$$

for Borel sets $B \subset \mathbb{R}^l$.

If we substitute $\mathbb{R}^l$ into the formula (17), we obtain Kac's formula, [18] Chapter IV, for the integral over $\mathbb{R}^l$ of the solution u of the problem (19).

Now the question arises, what happens if the (S,P,t)-measure does not exist? I do not know exactly, because, it is clear that a generalization of the concept of a σ-additive vector measure is needed for handling these type of problems. It may be easy to propose various generalizations, but the ultimate test of their viability will be whether we can use one of them to handle interesting cases such as wave propagation or quantum mechanical particles.

References

[1] W. Ambrose, *Spectral resolution of groups of unitary operators*, Duke Math. J. 11 (1944), 589-595.

[2] W.O. Amrein, V. Georgescu, and J.M. Jauch, *Stationary state scattering theory*, Helvetica Physica Acta 44 (1971), 407-434.

[3] R.G. Bartle, *A general bilinear vector integral*, Studia math. 15 (1956), 337-352.

[4] R.G. Bartle, N.S. Dunford and J.T. Schwartz, *Weak compactness and vector measures*, Canad. J. Math. 7 (1955), 289-305.

[5] Jürgen Batt, *A survey of some recent results on compact mappings*, Vector and operator valued measures and applications (Proc. Sympos., Snowbird Resort, Alta, Utah, 1972), pp. 23-32. Academic Press, New York, 1973.

[6] Jürgen Batt, *Die Verallgemeinerungen des Darstellumgssatzes von F. Riesz und ihre Anwendungen*, Jber. Deutsch. Math.-Verein. 74 (1973), 147-181.

[7] D. Blackwell, *The range of certain vector integrals*, Proc. Amer. Math. Soc. 2 (1951), 390-395.

[8] S.D. Chatterji, *Disintegration of measures and lifting*, Vector and operator valued measures and applications (Proc. Sympos., Snowbird Resort, Alta, Utah, 1972), pp. 69-83. Academic Press, New York, 1973.

[9] Ion Colojoară and Ciprian Foiaş, *Theory of Generalized Spectral Operators*, Mathematics and its Applications, Vol. 9, Gordon and Breach, New York-London-Paris, 1968.

[10] J. Diestel and J.J. Uhl, J.R., *Vector measures*, Mathematical Surveys No. 15, American Mathematical Society, Providence, Rhode Island, 1977.

[11] I. Dobrakov, *On integration in Banach spaces* I *and* II, Czechoslovak Math. J. 20 (95) (1970), 511-536, 680-695.

[12] Nelson Dunford and Jacob T. Schwartz, *Linear Operators, Part III, Spectral Operators*, Pure and Applied Mathematics, Vol. VII, Wiley-Interscience, New York, 1971.

[13] A. Dvoretzky, A. Wald, and J. Wolfowitz, *Relations among certain ranges of vector measures*, Pacific J. Math. 1 (1951), 59-74.

[14] E.B. Dynkin and I.V. Evstigneev, *Regular conditional expectations of correspondences*, Theor. Probability Appl. 21 (1976), 325-338.

[15] A. Goldman, *Mesures cylindriques, mesures vectorielles et questions de concentration cylindrique*, Pacific J. Math. 69 (1977), 385-413.

[16] L.L. Helms, *Introduction to potential theory*, Pure and Applied Mathematics Vol. XXII, Wiley-Interscience, New York, 1969.

[17] M. Jiřina, *Conditional probabilities on σ-algebras with countable basis*, Czechoslovak Math. J. 7 (1957), 130-153 (Russian; English translation in Selected Translations in Mathematical Statistics and Probability 2 (1962), 87-107.)

[18] M. Kac, *Probability and related topics in physical science* (Proc. Summer Seminar, Boulder, Colo., 1957), Interscience, New York, 1959.

[19] S. Karlin, *Extreme points of vector functions*, Proc. Amer. Math. Soc. 4 (1953), 603-610.

[20] J.F.C. Kingman and A.P. Robertson, *On a theorem of Lyapunov*, J. London Math. Soc. 43 (1968), 347-351.

[21] I. Kluvánek, *Characterization of Fourier-Stieltjes transforms of vector and operator valued measures*, Czechoslovak Math. J. 17 (92) (1967), 261-277.

[22] I. Kluvánek, *Fourier transforms of vector-valued functions and measures*, Studia Math. 37 (1970), 1-12.

[23] I. Kluvánek, *A compactness property of Fourier-Stieltjes transforms*, Matematický Časopis 20 (1970), 84-86.

[24] I. Kluvánek, Characterization of the closed convex hull of the range of a vector-valued measure, J. Functional Analysis 21 (1976), 316-329.

[25] I. Kluvánek, *Cylindrical measures and vector measures*, Commentationes Mathematicae Tomus Festivus (1978), 173-181.

[26] I. Kluvánek, *Operator-valued measures and perturbations of semi-groups*, Arch. Rational Mech. Anal. (to appear).

[27] Igor Kluvánek and Greg Knowles, *Vector measures and control systems*, Mathematics Studies 20, North-Holland, Amsterdam, 1975.

[28] Igor Kluvánek and Greg Knowles, *The bang-bang principle*, Mathematical Control Theory (Proceedings, Canberra, Australia, 1977), pp. 138-151, Lecture Notes in Mathematics 680, Springer-Verlag, Berlin, Heidelberg, New York, 1978.

[29] A. Kussmaul, *Stochastic integration and generalized martingales*, Research Notes in Maths, Collection π, Pitman Publishing, London, 1978.

[30] E.L. Lehman, *Testing statistical hypotheses*, Wiley, New York, 1959.

[31] D. Leviatan, *On a representation theorem and application to moment sequences in locally convex spaces*, Math. Ann. 182 (1969), 251-262.

[32] A. Liapunov, *Sur les fonctions-vecteurs complètement additives*, Izvestia Akad. Nauk SSSR Ser. Mat. 4 (1940), 465-478 (Russian, French summary).

[33] M. Loève, *Fonctions aléatoires du second ordre*, appendix to P. Lévy, *Processus stochastiques et mouvement brownien*, Gauthier-Villars, Paris, 1948.

[34] S.K. McKee, *Orthogonal expansions of vector-valued functions and measures*, Matematický Časopis 22 (1972), 71-80.

[35] S.K.McKee, *Transforms of vector-valued functions and measures*, Matematický Časopis 23 (1973), 5-13.

[36] E.J. McShane, *Integrals devised for special purposes*, Bull. Amer. Math. Soc. 69 (1963), 597-627.

[37] E.J. McShane, *Stochastic calculus and stochastic models*, Academic Press, New York, 1974.

[38] M. Marcus and V.J. Mizel, *A Radon-Nikodym type theorem for functionals*, J. Functional Analysis 23 (1976), 285-309.

[39] M. Marcus and V.J. Mizel, *Representation theorems for non-linear disjointly additive functionals and operators on Sobolev spaces*, Trans. Amer. Math. Soc. 228 (1977), 1-45.

[40] M. Métivier and J. Pellaumail, *Mesures stochastiques à valeurs dans des espaces* L_0, Z. Wahrscheinlichkeitstheorie verw. Gebiete 40 (1977), 101-114.

[41] A.G. Miamee and H. Salehi, *Harmonizability, V-boundedness and stationary dilation of stochastic process*, Indiana Univ. Math. J. 27 (1978), 37-50.

[42] N.W. Rickert, *The range of a measure*, Bull. Amer. Math. Soc. 73 (1967), 560-563.

[43] Paul Arthur Schilpp, editor, *Albert Einstein: Philosopher-Scientist*, The Library of Living Philosophers, Vol. VII, The Library of Living Philosophers, Inc., Evanston, Illinois, 1949.

[44] L. Schwartz, *Propriété de Radon-Nikodym*, Séminaire Maurey-Schwartz, Ecole Polytechnique, exposés no 4,5,6, année 1974-1975.

[45] Irving Segal, *Algebraic integration theory*, Bull. Amer. Math. Soc. 71 (1965), 419-489.

[46] Eric Thomas, *The Lebesgue-Nikodym theorem for vector valued Radon Measures*, Memoirs of the Amer. Math. Soc. 139, American Mathematical Society, Providence, R.I. 1974.

[47] Anthony K. Whitford, *Laplace-Stieltjes transforms of vector-valued measures*, Matematický Časopis 22 (1972), 156-163.

[48] Anthony K. Whitford, *Moment sequences in locally convex spaces*, Math. Ann. 205 (1973), 317-322.

[49] N. Wiener, *The homogeneous chaos*, Amer. J. Math. 60 (1938), 897-936.

[50] H.S. Witsenhausen, *A remark on reachable sets of linear systems*, IEEE Trans. Automatic Control 17 (1972), 547.

[51] Marc. Yor, *Quelques interactions entre mesures vectorielles et intégrales stochastiques*, Seminaire de Théorie du Potentiel (to appear in Springer, Lecture Notes in Mathematics).

[53] S. Zaidman, *Représentation des fonctions vectorielles par des intégrales de Laplace-Stieltjes et compacité faible*, C.r. Acad. Sci. Paris 248 (1959), 1915-1917.

[53] A.H. Zemanian, *Realizability theory for continuous linear system*, Mathematics in science and engineering 97, Academic Press, New York, 1972.

Contemporary Mathematics
Volume 2
1980

PETTIS'S MEASURABILITY THEOREM

J. J. Uhl, Jr.[1]

On the scale of theorem ratings most of us would rate Pettis's measurability theorem far below some of Pettis's other contributions. In fact if we would agree that the Dunford-Pettis Radon-Nikodym theorem is worth a glass of beer, then most of us would say that Pettis's measurability theorem is worth a glass of water. But it takes water to make beer and it takes Pettis's measurability theorem to prove the Dunford-Pettis Radon-Nikodym theorem. The fact is that many of the fundamental theorems in the theory of vector measures and the theory of Banach spaces rest squarely on Pettis's measurability theorem, and it is the purpose of this paper to demonstrate this point with specific examples.

On the other hand, Pettis's measurability theorem is also of no small general importance because it was (one of the) first theorems that demonstrated the depth to the weak topology of a Banach space. My view is that Pettis's measurability theorem, which Pettis proved in 1936, proves that Pettis was one of the very first mathematicians who understood that the weak topology can be important. A few years ago I asked Professor Pettis how he came to formulate a theorem now known as the Orlicz-Pettis theorem. He responded by saying that the measurability theorem involved the weak topology and a separability condition and that the O-P theorem (as he always called the Orlicz-Pettis theorem) also involves the weak topology and an implicit separability assumption.[2]

---

[1] supported in part by the National Science Foundation

[2] At another time, over a glass of wine, Professor Pettis asked me, "Jerry, what is the meaning of separability?" I gave him an answer which I was happy with at the time. Today I am embarrassed with my shallow answer and today I have no satisfactory answer to his question. On the other hand as a result of this interchange, today I know much more about separability than I did, and I relate this story becasue it typifies the way so many of us have learned from Professor Pettis.

Copyright © 1980, American Mathematical Society

Rather than rambling on I would like to turn to some specific applications of Pettis's measurability theorem, but before this can be done allow me to state the theorem.

Pettis's Measurability Theorem: Let (Ω,Σ,μ) be a finite measure space and X be a separable Banach space with dual X^*. A function $f: \Omega \longrightarrow X$ is the μ-almost everywhere limit of measurable simple functions (i.e. f is μ-measurable) if and only if the scalar valued function $\langle x^*,f\rangle$ is measurable for all x^* in a norming subset of X^*.

(Here Γ is a norming subset of X^* if $\|x\| = \sup_{x^*\in\Gamma} \langle \frac{x^*}{\|x^*\|}, x\rangle$ for all x in X.)

The proof of this theorem can be found in [1,II.1.2], in [3,III.6.11] or in Pettis's original paper [8] which appeared in 1938. By today's standards, the proof is not difficult, but I cannot believe that it was an easy theorem in 1936. Now let us look at some of the mathematical gems that rest on Pettis's measurability theorem.

1. The Radon-Nikodym property.

Let μ be Lebesgue measure on $[0,1]$. A Banach space X has the Radon-Nikodym property if every countably additive measure G from the Borel sets Σ in $[0,1]$ to X can be expressed in the form

$$G(E) = \lim_n \int_E g_n \, d\mu$$

where (g_n) is a sequence of simple functions with the property that

$$\lim_{n,m} \int_{[0,1]} \|g_n - g_m\|_X \, d\mu = 0.$$

As Pettis [9] and Dunford and Pettis [2] realized, if the set

$$\{G(E)/\mu(E): \ E \in \Sigma,\ \mu(E) > 0\}$$

has enough compactness properties (e.g. separable and relatively weakly compact or separable weak*-closed convex hull, then it is an easy matter [1, Chapter III] to find a separably-valued function $g: [0,1] \longrightarrow X$ such that

$$\langle x^*, G(E)\rangle = \int_E \langle x^*, g\rangle \, d\mu$$

for all E in Σ and all x^* in a norming subset Γ of X^*. Pettis's measurability theorem is immediately applicable to g and shows that g is the almost everywhere limit of a sequence (g_n) of simple functions. And with only slightly more work it is possible to show that

$$G(E) = \lim_n \int_E g_n \, d\mu$$

for all E in Σ and that

$$\lim_{m,n} \int_\Omega \|g_n - g_m\|_X \, d\mu = 0$$

It is this simple line of reasoning which shows that separable duals have the Radon-Nikodym property (Dunford-Pettis [2]) and that reflexive spaces have the Radon-Nikodym property (Dunford-Pettis [2] and Phillips [10]). It is not too much of an exaggeration to say that until 1978, with the appearance of Bourgain-Delbaen and McCarthy-O'Brien examples of separable Radon-Nikodym property spaces that do not embed in separable duals, every space with the Radon-Nikodym property had it as a result of an argument involving Pettis's measurability theorem.

There is a modern version of these arguments that has been profitably exploited by Saab [11,12,13]. It is an easy argument to prove that if

$$A = \{G(E)/\mu(E): E \in \Sigma, \mu(E) > 0\}$$

is contained in a norm separable weak*-compact convex subset of the dual of a separable space, then G admits the sort of representation discussed above. Saab has proved that if A is contained in a weak*-compact convex subset of a dual space on which the identity is weak*-to-weak universally Lusin measurable, then G admits the representation discussed above. In the course he characterized weak*-compact convex Radon-Nikodym subsets of a dual space. The lesson for us to learn is that he used universal measurability exactly the way Pettis's measurability theorem would be used in the separable case.

2. The Dunford-Pettis property for L_1.

A Banach space X has the Dunford-Pettis property if every weakly compact operator defined on X sends weakly convergent sequences onto norm convergent sequences. Ultimately it is Pettis's measurability theorem that guarantees that all L_1-spaces have the Dunford-Pettis property, a fact proved by Dunford and Pettis in their classic 1940 paper [2].

Here is the set-up. Let (Ω,Σ,μ) be an arbitrary measure space and let X be a Banach space. Let $T: L_1(\mu) \longrightarrow X$ be a weakly compact operator and suppose (f_n) is a weakly convergent sequence in $L_1(\mu)$. Let Σ_1 be the smallest sub σ-field of Σ relative to which all the f_n's are measurable and observe that we are in fact interested only in the action of $T: L_1(\Sigma_1,\mu) \longrightarrow X$. Thus without loss of generality μ is a σ-finite separable measure and hence without loss of generality μ is a finite measure and X is separable. The weak compactness of T and the separability of X allow us to use a compactness argument similar to the argument alluded to in the last section to find a $g: \Omega \longrightarrow X$ such that

$$\langle x^*, T(f)\rangle = \int_\Omega f\langle x^*, g\rangle \, d\mu$$

for all f in $L_1(\Sigma_1,\mu)$ and for all x^* in X^*. Since X is separable Pettis's measurability theorem applies and produces a sequence (g_n) of simple functions such that $\lim_n g_n = g$ almost everywhere. An appeal to Egorov's theorem reveals that for each $\delta > 0$ there is a set E_δ in Σ_1 with $\mu(\Omega \setminus E_\delta) < \delta$ such that $\lim_n g_n \chi_{E_\delta} = g\chi_{E_\delta}$ uniformly on Ω. A simple estimate shows that

$$\sup_{\|f\|\le 1} \left\| T(f\chi_{E_\delta}) - \int_{E_\delta} f g_n \, d\mu \right\|$$

tends to zero as n tends to infinity. This plus the fact that the operator

$$f \longrightarrow \int_{E_\delta} f g_n \, d\mu$$

is a finite rank operator (because each g_n has a finite range) shows that the operator

$$f \longrightarrow T(f\chi_{E_\delta})$$

is a compact operator.

Now to prove that the sequence $(T(f_n))$ is norm convergent, we shall prove the set $\{T(f_n): n = 1,2,3,...\}$ is totally bounded. To this end let $\varepsilon > 0$ and notice that since (f_n) is weakly convergent, then $\lim_n \int_E f_n \, d\mu$ exists for all E in Σ_1. Following Dunford, appeal to the Vitali-Hahn-Saks theorem [1,I.4.10] to see that

$$\lim_{\mu(E)\longrightarrow 0} \int_E |f_n| \, d\mu = 0$$

uniformly in n. Choose $\delta > 0$ such that

$$\int_E |f_n| \, d\mu < \varepsilon(2\|T\|)^{-1}$$

if $\mu(E) < \delta$ and choose E_δ in Σ_1 such that $\lim_n g_n \chi_{E_\delta} = g$ uniformly on E_δ and such that $\mu(\Omega \setminus E_\delta) < \delta$.

Then $T(f_n) = T(f_n\chi_{E_\delta}) + T(f_n \chi_{\Omega \setminus E_\delta})$. Since the operator $f \longrightarrow T(f\chi_{E_\delta})$ is compact, it is possible to cover $\{T(f_n \chi_{E_\delta}): n = 1,2,3,...\}$ with a finite number of $\varepsilon/2$ - balls. Since $\|T(f_n \chi_{\Omega \setminus E_\delta})\| < \varepsilon/2$ for all n, it follows that $\{T(f_n): n = 1,2,3,...\}$ can be covered with a finite number of ε-balls. This proves that $L_1(\mu)$ has the Dunford-Pettis property.

At this point it is very instructive to go over this proof. The proof depends crucially on Pettis's measurability theorem followed with a sweet application of Egorov's theorem

3. The Orlicz-Pettis Theorem.

The reason the Orlicz-Pettis theorem belongs in this paper is that the intuition Professor Pettis obtained from the measurability theorem led him to formulate his contribution to the Orlicz-Pettis theorem. After Professor Pettis told me this, I began to wonder whether the Orlicz-Pettis theorem can be deduced cleanly and simply from Pettis's measurability theorem. One day, as I was driving to Chapel Hill to join the Pettises for a couple of days at the beach, a proof dawned on me.

Orlicz-Pettis Theorem. A series in a Banach space is convergent (in norm) if each of its subseries is weakly convergent.

Proof. Let $\sum_{n=1}^{\infty} x_n$ be a formal series in a Banach space X and suppose that each of its subseries is weakly convergent. By the Hahn-Banach theorem, each weak limit of a subseries of $\sum_{n=1}^{\infty} x_n$ lies in the closed subspace of X spanned by the x_n's. Hence, without loss of generality, the space X is separable. (Note the analogy to the hypothesis of Pettis's Measurability Theorem.) Now suppose that $\sum_{n=1}^{\infty} x_n$ is not convergent and find and $\delta > 0$ and positive integers $m_1 < n_1 < m_2 < n_2 \cdots < m_j < n_j < \ldots$ such that if $y_j = \sum_{n=m_j}^{n_j} x_n$, then $\|y_j\| \geq \delta$. Note that the formal series $\sum_{j=1}^{\infty} y_j$ is a subseries of $\sum_{n=1}^{\infty} x_n$. Acoordingly each subseries of $\sum_{j=1}^{\infty} y_j$ is weakly convergent. Now let G be the Cantor group, i.e. the product $\{-1,1\}^N$ and μ be Haar measure on G, i.e. the coin tossing measure on G which is the product of the measure $\lambda(\{1\}) = \lambda(\{-1\}) = 1/2$ with itself countably many times. For each $\vec{\varepsilon} = (\varepsilon_1, \varepsilon_2, \ldots, \varepsilon_j, \ldots)$ in G let $f(\vec{\varepsilon})$ be the weak limit of the series $\sum_{j=1}^{\infty} \varepsilon_j y_j$. A moment's reflection shows that the function $f: G \longrightarrow X$ thus defined is continuous from G in its product topology to X in its weak topology. In particular $\langle x^*, f\rangle$ is measurable for all x^* in X^*. Thus f satisfies the hypothesis of Pettis's measurability theorem and is the μ-a.e. limit of simple functions.

Next it is important to realize that f is bounded. To see this define $T: X^* \longrightarrow \ell_1$ by $T(x^*) = \sum_{j=1}^{\infty} x^*(y_j)$ and apply the closed graph theorem. The upshot of this is that $f: G \longrightarrow X$ is μ-Bochner integrable, and hence

$$A = \{ \int_E f \, d\mu : E \subseteq G, \ E \text{ Borel}\}$$

is relatively compact in X [1,II.3.9].

But now if

$$E_n = \{\vec{\varepsilon} = \varepsilon_1, \varepsilon_2, \ldots, \varepsilon_n, \varepsilon_{n+1}, \ldots) \in G: \ \varepsilon_n = +1\}$$

then

$$x^* \int_{E_n} f \, d\mu = \int \ x^* f \, d\mu = \frac{x^*(y_n)}{2}$$

for all x^* in X^* and all $n = 1,2,3,\ldots$ Hence $\int_{E_n} f\,d\mu = \frac{y_n}{2}$ and the sequence (y_n) is relatively compact. Since $\sum_{n=1}^{\infty} y_n$ is weakly convergent, we know that $\lim_n y_n = 0$ weakly and hence in norm. This contradicts the fact that $\|y_n\| \geq \delta$ for all n. This completes the proof.

This proof can be generalized to include a small but important part of some recent theorems of Labuda [7] and Graves [4]. For an indication of what I mean, suppose Γ is a total subset of X^* and that the identity operator from X with its Γ topology to X with norm topology is universally Lusin measurable. It then follows that any series that is subseries convergent for the Γ topology of X is norm convergent. The proof goes as follows: Define f as in the above proof but use Γ-limits instead of weak limits. The assumption of universal measurability guarantees that f is μ-measurable. The Dieudonné-Grothendieck theorem [1,I.3.3] guarantees that f is bounded and hence Bochner integrable. The proof then proceeds just as above. I would like to go on and on about the Orlicz-Pettis theorem, but Professor Kalton will do a far better job in his paper than I would do here.

4. The Krein-Smulian Theorem.

Several basic theorems of Banach space theory owe their existence to Pettis's measurability theorem or arguments entirely analogous to its proof. The Krien-Smulian Theorem, the universal weak-to-norm Lusin measurability of the identity and the fact that regular measures defined on the weak Borel sets of a Banach space are nearly supported by norm compact sets are examples. In this section I will fight off the desire to talk about all of these things and be satisfied by discussing the Krein-Smulian Theorem. Watch for Pettis's measurability theorem.

Theorem(Krein-Smulian). *The norm-closed convex hull of a weakly compact set in a Banach space is weakly compact.*

Proof. Use Eberlien's theorem and a moment's reflection to reduce the proof to the separable case. Accordingly let W be a weakly compact subset of a separable Banach space X. Equip W with the weak topology and notice that the identity function f on W is (weakly) continuous and separably valved. By Pettis's measurability theorem, the identity f is μ-measurable for all μ in $(C(W,\text{weak}))^*$. Since f is obviously bounded, it follows that f is μ-Bochner integrable for every μ in $(C(W,\text{weak}))^*$. (Actually Pettis integrability is enough here.) Define an operator $T: (C(W,\text{weak}))^* \longrightarrow X$ by $T(\mu) = \int_W f\, d\mu$. Since the point masses are all in the unit ball of $C(W,\text{weak})^*$, it follows that

$$T(\text{Unit ball of}(C(W,\text{weak}))^*)$$

contains W. Moreover T is easily seen to be weak*-to-weakly continuous (Let (μ_α) be a net in $(C(W,\text{weak}))^*$ that converges weak* to some μ in $(C(W,\text{weak}))^*$. Take x^* in X^* and note that

$$\lim_\alpha \langle x^*, T(\mu_\alpha)\rangle = \lim_\alpha \int_W \langle x^*,f\rangle\, d\mu_\alpha$$

$$= \int_W \langle x^*,f\rangle\, d\mu = \langle x^*, T(\mu)\rangle$$

since $\langle x^*,f\rangle \in C(W,\text{weak})$ and $\lim_\alpha \mu_\alpha = \mu$ (weak*).) From this it follows that T maps the unit ball of $(C(W,\text{weak}))^*$ onto a weakly compact convex subset of X. Since this set contaims W, it contains the closed convex hull of W and we are done.

5. <u>Carleman integral operators</u>.

In their recent survey [6] Halmos and Sunder have rekindled interest in a subject that is dear to the heart of anyone who likes measure theory and functional analysis. Following them, let us agree that an operator $T: L_2[0,1] \longrightarrow L_2[0,1]$ is a Carleman integral operator if there exists a measureable real-valued function K on $[0,1] \times [0,1]$ such that

(1) $K(s,\cdot) \in L_2[0,1]$ for almost all s in $[0,1]$, and

(2) $Tf(s) = \int f(t)\, K(s,t)\, dt$

for amost all s in $[0,1]$ for all $f \in L_2[0,1]$.

The observations to make are that if we redefine K to be zero on the exceptional set in (1) and then define $g: [0,1] \longrightarrow L_2[0,1]$ by $g(s) = K(s,\cdot)$, then we see g is a vector valued function. Write $X = L_2[0,1]$ and notice that (2) says that $T(x^*) = \langle x^*,g\rangle$ for all x^* in $X^* = L_2[0,1]$. In particular $\langle x^*,g\rangle$ is measurable for all x^* in X^*. Since $X(=L_2[0,1])$ is separable, an appeal to

Pettis's Measurability Theorem reveals that g is the almost everywhere limit of a sequence (g_n) of simple X-valued functions. The result is a very strong compactness property of Carleman integral operators. Let $\varepsilon > 0$ and with the help of Egorov's theorem choose measurable set E such that $\lim_n g_n \chi_E = g\chi_E$ uniformly and such that the complement of E has measure less than ε.

Now notice that as the uniform limit of simple functions, the function $g\chi_E$ is bounded. Hence

$$\chi_E T(x^*) = \chi_E \langle x^*, g\rangle \in L_\infty[0,1]$$

for all x^* in $X^* = L_2[0,1]$. Not only does the operator

$$x^* \longrightarrow \chi_E T(x^*)$$

take $X^* = L_2[0,1]$ into $L_\infty[0,1]$ as a continuous operator; it does it also as a compact operator. To see why define $T_n: X^*(=L_2[0,1] \longrightarrow L_\infty[0,1]$ by $T_n(x^*) = \langle x^*, \chi_E\, g_n\rangle$. Since each g_n is a simple function, each T_n is a finite rank operator. Since $\lim_n \chi_E\, g_n = \chi_E g$ uniformly, we see

$$\lim_n \sup_{\|x^*\| \le 1} \| T_n(x^*) - \chi_E\, T(x^*) \|_{L_\infty}$$

$$= \lim_n \sup_{\|x^*\| \le 1} \| \langle x^*, \chi_E(g_n - g)\rangle \|_\infty$$

$$\le \lim_n \sup_{\|x^*\| \le 1} \|x^*\| \sup_{s \in E} \| g_n(s) - g(s) \|$$

$$= 0$$

since $\lim_n \chi_E\, g_n = \chi_E g$ uniformly on $[0,1]$. It follows that, as the operator topology limit of finite rank operators, the operator $x^* \longrightarrow \chi_E T(x^*)$ is compact from $L_2[0,1]$ to $L_\infty[0,1]$. This proves the following theorem:

Theorem. *If $T: L_2[0,1] \longrightarrow L_2[0,1]$ is a Carleman integral operator, then for each $\varepsilon > 0$ there is a measurable set E whose complement is of measure less than ε such that the operator $f \longrightarrow \chi_E T(f)$ from $L_2[0,1]$ to $L_\infty[0,1]$ is compact.*

This theorem as well as its converse and some of its generalizations can be found in [5].

I could go on and on about Pettis's Measurability theorem. But it is now time to stop and I think my point is made. Pettis's Measurability Theorem and its descendents are pervasive in operator theory, measure theory and pure Banach space theory. In the future let us recognize this fact and use it.

REFERENCES

1. J. Diestel and J. J. Uhl, Jr., Vector Measures. Math Survey no. 15 Amer. Math. Soc. Providence, 1977.
2. N. Dunford and B. J. Pettis, Linear operations on summable functions, Trans. Amer. Math. Soc. 47(1940), 323-392.
3. N. Dunford and J. T. Schwartz, Linear Operators Part I, Interscience, New York, 1958.
4. W. H. Graves, Universal Lusin measurability and subfamily summable families in Abelian topological groups. Proc. Amer. Math. Soc. 73(1979), 45-50.
5. N. Gretsky and J. J. Uhl, Jr., Careleman and Korotkov operators on Banach spaces (preprint).
6. P. R. Halmos and V. S. Sunder, Bounded Integral Operators on L^2 spaces, Springer-Verlag, Heidelberg, 1978.
7. I. Labuda, Universal measurability and summable families, (preprint).
8. B. J. Pettis, On integration in vector spaces, Trans. Amer. Math. Soc. 44(1938), 277-304.
9. B. J. Pettis, Differentiation in Banach spaces, Duke Math. J. 5(1939), 254-269.
10. R. S. Phillips, On linear transformations, Trans. Amer. Math. Soc 47(1940), 114-145.
11. E. Saab, Points extremaux, separabilité et faible K-analyticité dans les duaux d'espaces de Banach, C. R. Acad. Sci., Paris Sér. A, 285(1977), 1057-1060.
12. E. Saab, Une caracterisation des convexes $\sigma(E',E)$ compacts possédant la propriété de Radon-Nikodym, C. R. Acad. Sci., Paris Sér A, 286(1978), 45-48.
13. E. Saab, The Radon-Nikodym property in dual spaces: A theorem of Laurent Schwartz (to appear).

University of Illinois
Urbana, Illinois 61801

Contemporary Mathematics
Volume 2
1980

CLOSED MEASURES

by

Cecilia H. Brook and William H. Graves

A vector measure on an algebra of sets is closed if the algebra is a complete topological group in the Fréchet-Nikodým topology induced by the measure. Kluvánek introduced closed measures in [7]; the concept is central in the book of Kluvánek and Knowles [8]. According to [8], what makes closed measures useful is that their L^1 spaces are complete. Most measures encountered in applications are closed.

In this paper, using the universal measure theory set forth in [5] and [1], we consider closed measures as objects of interest in themselves. Roughly speaking, closed measures are those whose behavior is easily described in terms of the algebra of sets on which they are defined.

We characterize closedness in terms of continuity, orthogonality, and normality of vector measures. We show that the universal measure is closed precisely when the universal measure space, which is known to be the dual of a space of scalar measures, is just the space of bounded measurable functions. As corollary we obtain a theorem of Pettis involving real-valued measurable cardinals. We also prove a result of Kluvánek and Knowles characterizing closed measures in the absence of real-valued measurable cardinals.

Finally, we show that, on a σ-algebra, every measure satisfying the countable chain condition is closed. We use our results for closed measures to prove Drewnowski's theorem that a measure satisfies the countable chain condition iff it has a control measure.

Copyright © 1980, American Mathematical Society

1. Preliminaries

By locally convex space we shall mean a locally convex Hausdorff topological vector space over the complex field $\mathbb{C}$. If W and V are locally convex spaces, let $L(W,V)$ be the space of continuous linear maps from W to V. Denote $L(W,\mathbb{C})$ by W^*.

Let A be an algebra of subsets of a nonempty set X. Write $A_i \uparrow A$ to mean that (A_i) is an increasing sequence of sets in A with union A in A. Call (A_i) a disjoint sequence if it is a sequence of pairwise disjoint sets in A.

Let W be locally convex and ϕ an additive map from A to W. Say ϕ is strongly bounded (sb) if $\phi(A_i) \to 0$ whenever (A_i) is a disjoint sequence and countably additive (ca) if $\phi(A_i) \to \phi(A)$ whenever $A_i \uparrow A$. Call ϕ strongly countably additive (sca) if it is sb and ca. Let $ca(A,W)$ denote the space of ca maps from A to W and $sca(A,W)$ the space of sca maps from A to W. If $W = \mathbb{C}$, write just $ca(A)$ and $sca(A)$. Let $ca(A)^+$ be the space of nonnegative ca maps. Define $sca(A)^+$ similarly. We shall use the term measure for any set function which is at least finitely additive.

All results in this paper are derived by working in the universal measure space $L(A)$, which we now describe. For A an algebra, let $S(A)$ be the vector space of all complex-valued A-simple functions on X. Define the universal measure χ from A into $S(A)$ by letting $\chi(A) = \chi_A$, the characteristic function of A. For each additive map ϕ from A into W, define a linear map $\tilde{\phi}$ from $S(A)$ into W by letting $\tilde{\phi}(\chi_A) = \phi(A)$ and extending linearly. Of course, $\tilde{\phi}$ is just integration with respect to ϕ. Let the universal measure topology τ be the weakest locally convex topology on $S(A)$ making $\tilde{\phi}$ continuous for every locally convex W and every ϕ in $sca(A,W)$.

Now let $L(A)$ be the completion of $S(A)$ with respect to τ. For ϕ in $sca(A,W)$ where W is complete locally convex, let $\hat{\phi}$ denote the unique continuous linear extension of $\tilde{\phi}$ to L(A). Then χ from A into L(A) is sca [5, Lemma 1.3] and, for every complete locally convex W, the map $\phi \to \hat{\phi}$ is a linear bijection of sca(A,W) onto L(L(A), W) [5, Theorem 1.5]. A deeper result is that L(A) is the dual of sca(A) relative to the total variation norm [5, Theorem 10.5].

Last, L(A) is a commutative C^*-algebra. For with the sup norm, pointwise multiplication, and complex conjugation, S(A) is a commutative normed *-algebra with identity. By viewing L(A) as the norm-dual of sca(A), which in turn is the τ-dual of S(A), we may define an Arens' multiplication and a natural involution on the Banach space L(A).

1.1 Theorem. (1) L(A) is a commutative C^*-algebra with identity in which S(A) is isometrically *-embedded.

(2) The universal measure topology τ is weaker than the norm topology on L(A).

(3) Multiplication is separately τ-continuous and jointly τ-continuous on norm-bounded sets in L(A).

Proof. For (1) see [1, Theorem 2.24]. (2) follows from [5, Theorem 2.1], and (3) is [1, Theorem 2.12].

Call P in L(A) a projection if $P^2 = P$ and $P^* = P$. Clearly every characteristic function is a projection. The set P of all projections in L(A) carries a natural order: $Q \le P$ if $PQ = Q$.

1.2 Theorem. (1) The set P is a complete Boolean algebra.

(2) Every increasing net in P τ-converges to its least upper bound.

(3) The τ-closure of $\chi(A)$ in L(A) is P. Moreover, any projection can be reached from $\chi(A)$ by taking four successive τ-limits of monotone (increasing or decreasing) nets.

(4) There is a neighborhood base for τ on L(A) consisting of absolutely convex sets U with the property that if P and Q are projections with $Q \le P$ and P is in U, then Q is in U.

Proof. See [1, Theorems 3.16, 3.13, 5.26-27, and 6.2].

In [1] projections are used to study continuity properties of measures. Let ϕ be in sca(A,W) and ψ in sca(A,V) with W, V locally convex. For E in A let $A_E = \{A \in A | A \subseteq E\}$. Say ψ is ϕ-continuous if for every neighborhood (of zero) M in V there is a neighborhood N in W such that $\psi(A_E) \subseteq M$ whenever $\phi(A_E) \subseteq N$. Say ψ is algebraically ϕ-continuous, and write $\psi << \phi$, if $\Delta_\phi \subseteq \Delta_\psi$, where Δ_ϕ is the ideal of all E in A with $\phi(A_E) = 0$. Call ϕ and ψ algebraically orthogonal if there is E in A such that E is in Δ_ϕ and X - E in Δ_ψ, and topologically orthogonal if for every pair of neighborhoods N in W and M in V there is E in A with $\phi(A_E) \subseteq N$ and $\psi(A_{X-E}) \subseteq M$. Clearly, continuity implies algebraic continuity, and algebraic orthogonality implies topological orthogonality.

For P in P let $P(A) = P\chi_A$. By 1.1(3), $P : A \to (L(A), \tau)$ is sca. Thus every projection defines a measure.

Fix ϕ in sca(A,W) with W complete locally convex. For P in P, let $(\phi \circ P)(A) = \hat{\phi}(P\chi_A)$. Let $P_\phi = \wedge \{P \in P | \phi \circ P = \phi\}$. Then P_ϕ is a projection which may be thought of as the support of ϕ. For Q in P, clearly $P_Q = Q$.

1.3 Theorem. (1) $\phi = \phi \circ P_\phi$.

(2) $P_\phi \chi_A = 0$ iff A is in Δ_ϕ.

(3) ϕ is P_ϕ-continuous and P_ϕ is ϕ-continuous.

Proof. See [1, Theorems 4.2, 4.18, 6.24].

Two measures may be compared by comparing their projections.

1.4 Theorem. Let ϕ be in sca(A,W) and ψ in sca(A,V) with W,V complete locally convex. Then

(1) ψ is ϕ-continuous iff $P_\psi \leq P_\phi$.

(2) ψ and ϕ are topologically orthogonal iff $P_\psi P_\phi = 0$.

Proof. See [1, Theorem 6.26].

Projections corresponding to nonnegative measures play a special role. Let $L = \{P_\mu | \mu \in \text{sca}(A)^+\}$.

1.5 Theorem. The set L is a dense σ-ideal in P. That is,

(1) For every $P \neq 0$ there is $Q \neq 0$ such that $Q \leq P$ and Q is in L.

(2) If $P \leq Q$ for some Q in L, then P is in L.

(3) If $P = \vee Q_n$ with all Q_n in L, then P is in L.

Proof. See [1, Theorem 4.25].

2. Closed Measures

In this section we define closed measure and see how closedness of ϕ is reflected in the projection P_ϕ. We also look at the relationship between completeness of A/Δ_ϕ as a topological group and as a Boolean algebra, and give our first characterization of closed measures.

Let A be an algebra and fix ϕ in $\mathrm{sca}(A,W)$ with W complete locally convex. It follows from 1.2(1) that $P_\phi P = \{Q \in P | Q \leq P_\phi\}$ is a complete Boolean algebra. For P and Q in P, let $P \Delta Q = P(I-Q) + Q(I-P)$, the symmetric difference of P and Q. By 1.1(3) $P_\phi P$ is an abelian topological group under Δ, containing $P_\phi A = \{P_\phi \chi_A | A \in A\}$ as a subgroup.

Define the ϕ-topology (denoted by G_ϕ) on A to be the topology with neighborhood base at B consisting of all sets $\{E | \phi(A_{E\Delta B}) \subseteq N\}$ where N is an absolutely convex neighborhood in W. Then (A, G_ϕ) is an abelian topological group under the operation of symmetric difference. Note that G_ϕ is a Fréchet-Nikodým topology in the sense of Drewnowski [3]. Since the closure of $\{\emptyset\}$ in the ϕ-topology is Δ_ϕ, G_ϕ is Hausdorff iff $\Delta_\phi = \{\emptyset\}$. Let G_ϕ also denote the quotient topology induced on A/Δ_ϕ by the ϕ-topology on A. Then $(A/\Delta_\phi, G_\phi)$ is a Hausdorff abelian topological group, with neighborhood base at $\overline{B}$ consisting of all $\{\overline{E} | \phi(A_{E\Delta B}) \subseteq N\}$ where N is an absolutely convex neighborhood in W.

2.1 Lemma. (1) The map $\overline{A} \to P_\phi \chi_A$ of $(A/\Delta_\phi, G_\phi)$ onto $(P_\phi A, \tau)$ is an isomorphism of topological groups.

(2) The set $P_\phi A$ is τ-complete in $L(A)$ iff $P_\phi A = P_\phi P$.

Proof. (1) The map is well-defined by 1.3(2). To see that it is a homeomorphism, use 1.3(3) and 1.2(4).

(2) By 1.1(3) and 1.2(3) the τ-closure of $P_\phi A$ is $P_\phi P$. The result follows.

The measure ϕ is <u>closed</u> if (A, G_ϕ) is a complete topological group.

2.2 <u>Theorem</u>. The following are equivalent:

(1) ϕ is closed.

(2) $(A/\Delta_\phi, G_\phi)$ is a complete topological group.

(3) $P_\phi A = P_\phi P$.

Proof. That (1) and (2) are equivalent is straightforward. (2) and (3) are equivalent by 2.1.

We next state some results which provide examples of closed measures.

2.3 <u>Theorem</u>. If A is a σ-algebra and W a Fréchet space, then every ϕ in $ca(A,W)$ is closed.

Proof. See [8, Theorem IV. 7.1].

For each x in X, let δ_x be the 0-1 valued point mass measure at x. Then δ_x is in $sca(A)^+$. Let $P_X = \vee \{P_{\delta_x} | x \in X\}$, and let $I = \chi_X$, the identity of $L(A)$. Recall that card X (the cardinal number of X) is <u>real-valued measurable</u> if there is μ in $ca(2^X)^+$ such that $\mu(\{x\}) = 0$ for each x in X but $\mu(X) = 1$.

2.4 <u>Theorem</u>. (1) The projection P_X is a closed measure iff $A = 2^X$.
(2) Let $A = 2^X$. Then $P_X \neq I$ iff card X is real-valued measurable.

Proof. See [2, Theorems 3.9, 3.10].

Now A/Δ_ϕ is a Boolean algebra as well as a topological group, and the map $\bar{A} \to P_\phi \chi_A$ of A/Δ_ϕ into $P_\phi P$ is a Boolean isomorphism (it preserves finite operations).

2.5 Lemma. If ϕ is closed, then A/Δ_ϕ is a complete Boolean algebra and the map $\bar{A} \to P_\phi \chi_A$ is a complete isomorphism of A/Δ_ϕ onto $P_\phi \mathcal{P}$.

Proof. By 2.2 $P_\phi A = P_\phi \mathcal{P}$. Since $P_\phi \mathcal{P}$ is a complete Boolean algebra, the rest follows.

See 4.1 for a case in which A/Δ_ϕ is complete but ϕ is not closed.

If ψ is in sca(A,V) with V locally convex and if $\psi << \phi$, define $\bar{\psi}$ on A/Δ_ϕ by $\bar{\psi}(\bar{A}) = \psi(A)$. Then $\bar{\psi}$ is well-defined and additive. If P is a projection vanishing on Δ_ϕ, write P for $\bar{P}$.

If $\mathcal{B}$ is a Boolean algebra and λ an additive map from $\mathcal{B}$ to W, say λ is normal if $\lambda(B_\alpha) \to \lambda(B)$ whenever (B_α) is an increasing net in $\mathcal{B}$ with least upper bound B.

2.6 Lemma. If A/Δ_ϕ is a complete Boolean algebra and P_ϕ is normal on A/Δ_ϕ, then ϕ is closed.

Proof. First notice that if $(P_\phi \chi_{A_\alpha})$ is an increasing or decreasing net in $P_\phi A$ with τ-limit Q, then $Q = P_\phi \chi_A$ for some A in A. For if $(P_\phi \chi_{A_\alpha})$ is increasing, then $(\bar{A}_\alpha)$ is increasing in A/Δ_ϕ. Let $\bar{A} = \vee \bar{A}_\alpha$. Since P_ϕ is normal on A/Δ_ϕ, $P_\phi \chi_{A_\alpha} \to P_\phi \chi_A$ τ. Then $Q = P_\phi \chi_A$. The argument for a decreasing net is similar.

Now apply 1.1(3) and 1.2(3) to conclude that $P_\phi A = P_\phi \mathcal{P}$. By 2.2 ϕ is closed.

2.7 Theorem. Let A be an algebra, ϕ in sca(A,W), ψ in sca(A,V), with W, V complete locally convex. Then ϕ is closed iff A/Δ_ϕ is a complete Boolean algebra on which $\bar{\psi}$ is normal whenever ψ is ϕ-continuous.

Proof. Suppose that ϕ is closed. It follows from 2.5 that A/Δ_ϕ is complete and P_ϕ is normal on A/Δ_ϕ. If ψ is ϕ-continuous, then by 1.4(1) $P_\psi \le P_\phi$. By 1.1(3) P_ψ is normal on A/Δ_ϕ. By 1.3(1) $\bar{\psi}$ is normal on A/Δ_ϕ.

For the converse, use 1.3(3) and 2.6.

Kluvánek and Knowles show that if ϕ is closed, then A/Δ_ϕ is complete [8, Theorem IV. 5.1].

3. Continuity and Orthogonality

In this section we characterize closed measures in terms of continuity and orthogonality.

If Q is a projection in $L(A)$, then $Q(A_E) = 0$ iff $Q\chi_E = 0$. It follows that Q and R in P are algebraically orthogonal iff $Q\chi_A = 0$ and $R(I-\chi_A) = 0$ for some A in A. Then Q is algebraically orthogonal to every subprojection of R whenever Q and R are algebraically orthogonal. By 1.4(2) Q and R are topologically orthogonal iff $QR = 0$.

3.1 Lemma. For Q and P in P with $Q \le P$, Q is algebraically orthogonal to $P - Q$ iff $Q = P\chi_B$ for some B in A.

Proof. If $Q = P\chi_B$, then $P - Q = P(I-\chi_B)$. Let $A = X - B$. Then $Q\chi_A = 0$ and $(P-Q)(I-\chi_A) = 0$.

Conversely, if $Q\chi_A = 0$ and $(P-Q)(I-\chi_A) = 0$, let $B = X - A$. Then $P\chi_B = P(I-\chi_A) = (P-Q)(I-\chi_A) + Q(I-\chi_A) = Q$.

3.2 Theorem. Let A be an algebra, ϕ in $sca(A,W)$, ψ_1 in $sca(A, V_1)$ and ψ_2 in $sca(A, V_2)$, with W, V_1, V_2 complete locally convex. Then ϕ is closed iff ψ_1 and ψ_2 are algebraically orthogonal whenever ψ_1 and ψ_2 are ϕ-continuous and topologically orthogonal.

Proof. Suppose that ϕ is closed and ψ_1 and ψ_2 are ϕ-continuous and topologically orthogonal. By 1.4 $P_{\psi_1} \le P_\phi$, $P_{\psi_2} \le P_\phi$, and $P_{\psi_1} P_{\psi_2} = 0$. By 2.2 $P_{\psi_1} = P_\phi\chi_B$ for some B in A. By 3.1 P_{ψ_1} and $P_\phi - P_{\psi_1}$, hence P_{ψ_1} and P_{ψ_2}, are algebraically orthogonal. By 1.3(2) ψ_1 and ψ_2 are algebraically orthogonal.

Conversely, let $Q \le P_\phi$. By 1.4 Q and $P_\phi - Q$ are topologically orthogonal and ϕ-continuous, hence algebraically orthogonal. By 3.1 $Q = P_\phi\chi_B$ for some B in A. By 2.2 ϕ is closed.

3.3 Lemma. Let P be a projection in $L(A)$. If $P = Q$ whenever Q is in PP and $P << Q$ as measures, then for every nonzero R in PP there is A in A such that $P\chi_A \neq 0$ and $P\chi_A \leq R$.

Proof. There is A in A such that $(P-R)\chi_A = 0$ but $P\chi_A \neq 0$. For otherwise $P << P - R$, which implies that $P = P - R$. Since $(P-R)\chi_A = 0$, $P\chi_A = R\chi_A \leq R$.

3.4 Theorem. Let A be an algebra, ϕ in $sca(A,W)$, ψ_1 in $sca(A, V_1)$ and ψ_2 in $sca(A, V_2)$, with W, V_1, V_2 complete locally convex. Then ϕ is closed iff A/Δ_ϕ is a complete Boolean algebra and ψ_1 is ψ_2-continuous whenever ψ_1 and ψ_2 are ϕ-continuous and $\psi_1 << \psi_2$.

Proof. Suppose that ϕ is closed. By 2.5 A/Δ_ϕ is complete. Let ψ_1 and ψ_2 be ϕ-continuous and $\psi_1 << \psi_2$. By 1.4(1) $P_{\psi_1} \leq P_\phi$ and $P_{\psi_2} \leq P_\phi$. By 2.2 $P_{\psi_2} = P_\phi\chi_B$ for some B in A. Let $A = X - B$. Then $P_{\psi_2}\chi_A = 0$, whence $P_{\psi_1}\chi_A = 0$ by 1.3(2). Then $P_{\psi_1} = P_{\psi_1} - P_{\psi_1}\chi_A = P_{\psi_1}\chi_B = P_{\psi_1}P_\phi\chi_B = P_{\psi_1}P_{\psi_2}$, so $P_{\psi_1} \leq P_{\psi_2}$. By 1.4(1) ψ_1 is ψ_2-continuous.

For the converse, first note that the conditions of 3.3 are satisfied. Now choose Q in $P_\phi P$ with $Q \neq 0$ and $Q \neq P_\phi$. Let $B = \{A \in A | P_\phi\chi_A \leq Q\}$. Let $\overline{B} = \vee\{\overline{A} \in A/\Delta_\phi | A \in B\}$, which exists since A/Δ_ϕ is complete. Let $Q' = \vee\{P_\phi\chi_A | A \in B\}$. Clearly $Q' \leq Q$ and $Q' \leq P_\phi\chi_B$. It follows easily from 3.3 that $Q' = Q$. If $Q' \neq P_\phi\chi_B$, then by 3.3 there is C in A such that $P_\phi\chi_C \neq 0$ and $P_\phi\chi_C \leq P_\phi\chi_B - Q'$. Then $Q' \leq P_\phi\chi_B - P_\phi\chi_C = P_\phi\chi_{B-C} \lneq P_\phi\chi_B$, so, for all A in B, $\overline{A} \leq \overline{B - C}$ and $\overline{B - C} \lneq \overline{B}$. This contradiction shows that $Q' = P_\phi\chi_B$. Then $Q = P_\phi\chi_B$. By 2.2 ϕ is closed.

After giving one more definition, we summarize our results thus far.

Let $BM(A)$ denote the norm-closure of $S(A)$ in $L(A)$. If A is a σ-algebra, then $BM(A)$ is (isometrically *-isomorphic to) the space of bounded A-measurable complex-valued functions on X.

3.5 Theorem. Let A be an algebra, ϕ in $sca(A,W)$, ψ_1 in $sca(A,V_1)$, and ψ_2 in $sca(A,V_2)$ with W, V_1, V_2 complete locally convex. Then the following are equivalent:

(1) ϕ is closed.

(2) $P_\phi A = P_\phi P$.

(3) $P_\phi\, BM(A) = P_\phi\, L(A)$.

(4) A/Δ_ϕ is a complete Boolean algebra on which $\overline{\psi}_1$ is normal whenever ψ_1 is ϕ-continuous.

(5) A/Δ_ϕ is a complete Boolean algebra and ψ_1 is ψ_2-continuous whenever ψ_1 and ψ_2 are ϕ-continuous and $\psi_1 << \psi_2$.

(6) ψ_1 and ψ_2 are algebraically orthogonal whenever ψ_1 and ψ_2 are ϕ-continuous and topologically orthogonal.

Proof. By [2, Theorem 3.5] (2) and (3) are equivalent. The rest has been proved.

Note that if we take ϕ to be the universal measure χ, then, since $P_\chi = I$, 3.5 tells us when the universal measure space $L(A)$ is just $BM(A)$. We shall see in the next section that more can be said.

4. The Role of Real-valued Measurable Cardinals

Here set theory enters the picture. We show that, in the absence of real-valued measurable cardinals, the universal measure on A is closed iff $A = 2^X$ and, in general, a measure ϕ is closed iff A/Δ_ϕ is a complete Boolean algebra. (The second result is proved by Kluvánek and Knowles [8, Theorem IV. 5.1].)

Let B be a Boolean algebra and λ an additive map from B to W. Call λ completely additive if whenever (C_α) is a disjoint family in B with least upper bound C, then $(\lambda(C_\alpha))$ is summable in W with sum $\lambda(C)$. If λ is normal, then λ must be completely additive.

4.1 Theorem. Let A be an algebra. The following are equivalent:

(1) χ is closed, or equivalently $\chi(A) = P$.

(2) $L(A) = BM(A)$.

(3) $A = 2^X$ and, for every complete locally convex W, every ϕ in $ca(2^X, W)$ is completely additive on 2^X.

(4) $A = 2^X$ and every μ in $ca(2^X)^+$ is completely additive on 2^X.

(5) $A = 2^X$ and card X is not real-valued measurable.

Proof. By 3.5, (1) and (2) are equivalent.

(1) implies (3): By 3.5 A is complete and every ϕ in $sca(A,W)$ is normal, hence completely additive, on A. To see that $A = 2^X$, it is enough to show that A contains all singletons. But given x in X there is A in A such that $P_{\delta_x} = \chi_A$, where P_{δ_x} is the point mass measure at x. Now 1.3(2) implies that $A = \{x\}$.

(3) implies (4) is obvious.

(4) implies (5): If μ is in $ca(2^X)^+$ and $\mu(\{x\}) = 0$ for all x in X, then, since μ is completely additive, $\mu = 0$.

(5) implies (1): By 2.4 the identity projection I is closed, so $\chi(A) = P$.

In [9] Pettis gives some measure-theoretic consequences of the hypothesis that card X is not real-valued measurable. If we start with $A = 2^X$, then the equivalence of (2) and (5) in 4.1 is precisely [9, Theorem 2.3].

We now give an example promised earlier.

4.2 Example. If $A = 2^X$ and card X is real-valued measurable, then $A/\Delta_\chi = A$ is a complete Boolean algebra, but by 4.1 χ is not closed.

Next we give another proof of [8, Theorem IV. 5.1], which says that if there are no real-valued measurable cardinals, then A/Δ_ϕ is complete as a topological group iff it is complete as a Boolean algebra.

4.3 Lemma. Let B be a Boolean algebra and λ an additive map on B which is nonnegative or projection-valued. Then λ is completely additive iff λ is normal.

Proof. Suppose λ is completely additive. Let (B_α) be an increasing net in B with least upper bound B. By Zorn's lemma, there is a disjoint family (C_β) in $\mathcal{B}$ satisfying

(1) For each β there is α such that $C_\beta \subseteq B_\alpha$,

(2) $\vee C_\beta = B$.

Then $\Sigma\, \lambda(C_\beta) = \lambda(B)$. Now use the order properties of range λ to show that $\lambda(B_\alpha) \to \lambda(B)$. Then λ is normal. The converse is clear.

4.4 Lemma. Let A be a σ-algebra and P a projection in $L(A)$. If A/Δ_P is a complete Boolean algebra and if A/Δ_P contains no disjoint family of real-valued measurable cardinal, then P is completely additive on A/Δ_P.

Proof. Suppose that A/Δ_P is complete but P is not completely additive on A/Δ_P. Then there is a disjoint family $(\overline{C}_\alpha)_{\alpha \in J}$ with $\overline{C} = \vee \overline{C}_\alpha$ but $P\chi_C \neq \Sigma\, P\chi_{C_\alpha}$ in $L(A)$. Now $(P\chi_{C_\alpha})$ is a disjoint family in P. Let $Q = \Sigma\, P\chi_{C_\alpha} = \vee\, P\chi_{C_\alpha}$. Then $Q \lneq P\chi_C$. By 1.5(1) there is μ in $ca(A)^+$ such that $P_\mu \neq 0$ and $P_\mu \leq P\chi_C - Q$. Then $P_\mu Q = 0$ and $P_\mu \leq P\chi_C \leq P$. For each α in J, $P_\mu \chi_{C_\alpha} = P_\mu P\chi_{C_\alpha} = P_\mu Q\chi_{C_\alpha} = 0$, while $P_\mu \chi_C = P_\mu P\chi_C = P_\mu \neq 0$. By 1.3(2) $\mu(C_\alpha) = 0$ for all α in J but $\mu(C) \neq 0$.

Since $P_\mu \leq P$, by 1.3(2) $\overline{\mu}$ is well-defined on A/Δ_P. Since A is a σ-algebra, $\overline{\mu}$ is ca.

Define ν on 2^J by letting $\nu(S) = \overline{\mu}(\vee_{\alpha \in S} \overline{C}_\alpha)$ for each subset S of J. Since $\overline{\mu}$ is ca, so is ν. For each α in J, $\nu(\{\alpha\}) = \mu(C_\alpha) = 0$, but $\nu(J) = \mu(C) \neq 0$. Then card J is real-valued measurable. Thus A/Δ_P contains a disjoint family of real-valued measurable cardinal.

4.5 Theorem. Let A be a σ-algebra and ϕ in $ca(A,W)$, W complete locally convex. If A/Δ_ϕ contains no disjoint family of real-valued measurable cardinal, then ϕ is closed iff A/Δ_ϕ is a complete Boolean algebra.

Proof. If ϕ is closed, then A/Δ_ϕ is complete by 2.5.

If A/Δ_ϕ is complete, then by 1.3(2) and 4.4, P_ϕ is completely additive on A/Δ_ϕ. By 4.3 P_ϕ is normal on A/Δ_ϕ. Now by 2.6 ϕ is closed.

Theorem 4.5 has a curious consequence concerning continuity of measures. To derive it, we use another projection associated with ϕ [5, §7]. Let $P_{\Delta_\phi} = \vee\{\chi_A \mid A \in \Delta_\phi\}$.

4.6 <u>Lemma</u>. (1) $P_\phi \le I - P_{\Delta_\phi}$.

(2) $(I-P_{\Delta_\phi})\,\chi_A = 0$ iff A is in Δ_ϕ.

(3) $P_\phi = I - P_{\Delta_\phi}$ iff ψ is ϕ-continuous whenever $\psi << \phi$ and ψ is in $sca(\mathcal{A},V)$ with V complete locally convex.

Proof. For (1) see [1, Theorem 4.18]. (2) is straightforward, and (3) is [1, Theorem 6.29].

4.7 <u>Theorem</u>. Let $\mathcal{A}$ be a σ-algebra, ϕ in $ca(\mathcal{A},W)$, ψ in $ca(\mathcal{A},V)$ with W, V complete locally convex. If ϕ is closed and $\mathcal{A}/\Delta_\phi$ contains no disjoint family of real-valued measurable cardinal, then ψ is ϕ-continuous whenever $\psi << \phi$.

Proof. Let $Q = I - P_{\Delta_\phi}$. By 4.6(2), $\Delta_Q = \Delta_\phi$. By 4.5 applied to ϕ, $\mathcal{A}/\Delta_Q$ is complete. By 4.5 applied to Q, Q is closed. By 4.6(1), Q and P_ϕ are Q-continuous. By 4.6(2) $Q << P_\phi$. Now by 3.4, Q is P_ϕ-continuous, or $Q \le P_\phi$. Then $Q = P_\phi$. By 4.6(3) the result follows.

5. The Countable Chain Condition

In this section we show that, on a σ-algebra, every measure satisfying the countable chain condition is closed. We use this result to give a new and perhaps simpler proof of a theorem of Drewnowski.

Throughout this section let $\mathcal{A}$ be a σ-algebra. An additive map λ on $\mathcal{A}$ satisfies the <u>countable chain condition</u> (ccc) if whenever $(A_\alpha)_{\alpha\in I}$ is a disjoint family in $\mathcal{A}$ and $\lambda(A_\alpha) \ne 0$ for all α in I, then I is countable. A Boolean algebra $\mathcal{B}$ satisfies ccc if every disjoint family of nonzero elements of $\mathcal{B}$ is countable.

5.1 <u>Lemma</u>. Let P be a projection which satisfies ccc as a measure. If $(A_\alpha)_{\alpha\in I}$ is a family in $\mathcal{A}$ for which $(P\chi_{A_\alpha})_{\alpha\in I}$ is a disjoint family of nonzero projections, then I is countable and there is a disjoint family $(B_\alpha)_{\alpha \in I}$ in A such that $P\chi_{B_\alpha} = P\chi_{A_\alpha}$ for all α in I.

Proof. Let C be the collection of all pairs $(J, (C_\alpha)_{\alpha \in J})$ such that $J \subseteq I$, $(C_\alpha)_{\alpha \in J}$ is a disjoint family in A, and $P\chi_{C_\alpha} = P\chi_{A_\alpha}$ for all α in J. Clearly C is nonempty. Order C by letting $(J_1, (C_\alpha)_{\alpha \in J_1}) < (J_2, (D_\alpha)_{\alpha \in J_2})$ if $J_1 \subseteq J_2$ and $C_\alpha = D_\alpha$ for all α in J_1. By Zorn's lemma C contains a maximal element $(J, (B_\alpha)_{\alpha \in J})$. Since $P\chi_{B_\alpha} = P\chi_{A_\alpha} \neq 0$ for all α in J, J is countable. If $J \neq I$, find α_0 in $I - J$. Let $B = \bigcup_{\alpha \in J} B_\alpha$ and $B_{\alpha_0} = A_{\alpha_0} - B$. Then B_{α_0} is in A and B_{α_0} misses B_α for all α in J. Since $P\chi_{A_{\alpha_0}} \chi_B = P\chi_{A_{\alpha_0}} (\vee \chi_{B_\alpha}) = \vee(P\chi_{A_{\alpha_0}} \chi_{B_\alpha}) = \vee(P\chi_{A_{\alpha_0}} P\chi_{A_\alpha}) = 0$, $P\chi_{B_{\alpha_0}} = P\chi_{A_{\alpha_0}}$. Let $J_0 = J \cup \{\alpha_0\}$. Then $(J_0, (B_\alpha)_{\alpha \in J_0})$ is an element of C strictly greater than $(J, (B_\alpha)_{\alpha \in J})$, which is impossible. Therefore $I = J$.

5.2 Lemma. The following are equivalent:

(1) ϕ satisfies ccc.

(2) P_ϕ satisfies ccc.

(3) A/Δ_ϕ satisfies ccc.

Proof. (1) and (2) are equivalent by 1.3(2). By 5.1 (2) implies (3). Clearly (3) implies (2).

5.3 Theorem. Let A be a σ-algebra and ϕ in $ca(A,W)$, W complete locally convex. If ϕ satisfies ccc, then ϕ is closed.

Proof. Since A is a σ-algebra, A/Δ_ϕ is σ-complete. By 5.2 A/Δ_ϕ satisfies ccc. Therefore (see [6, §14]) A/Δ_ϕ is complete. Again by 5.2 A/Δ_ϕ contains no disjoint family of real-valued measurable cardinal. By 4.5 ϕ is closed.

5.4 Theorem. Let A be a σ-algebra and W a Fréchet space. If ϕ is in $ca(A,W)$, then ϕ satisfies ccc and hence is closed.

Proof. That ϕ satisfies ccc follows from the metrizability of W and the strong boundedness of ϕ. By 5.3 ϕ is closed.

Observe that we have now proved 2.3 without the use of control measures. Also note that for ϕ as in 5.4 the conclusion of 4.7 holds.

5.5 Theorem. For P in $\mathcal{P}$, the following are equivalent:

(1) $P = P_\mu$ for some μ in $ca(\mathcal{A})^+$.

(2) The measure P satisfies ccc.

(3) The Boolean algebra $P\mathcal{P}$ satisfies ccc.

Proof. By 5.2 and 5.4, (1) implies (2).

(2) implies (3): By 5.3 P is closed. By 2.2 $P\mathcal{P} = P\mathcal{A}$. But $P\mathcal{A}$ and $\mathcal{A}/\Delta_P$ are isomorphic Boolean algebras. By 5.2 $\mathcal{A}/\Delta_P$ satisfies ccc.

(3) implies (1): Using Zorn's lemma, find a maximal disjoint family of nonzero subprojections of P which are of the form P_μ. Since this family must be countable, we may write it (Q_n). By 1.5(1) $P = \vee Q_n$. By 1.5(3), $P = P_\mu$ for some μ in $ca(\mathcal{A})^+$.

Call μ in $ca(\mathcal{A})^+$ a control measure for ϕ if ϕ is μ-continuous.

5.6 Theorem. Let $\mathcal{A}$ be a σ-algebra. For ϕ in $ca(\mathcal{A},W)$, W complete locally convex, the following are equivalent:

(1) ϕ satisfies ccc.

(2) P_ϕ satisfies ccc as a measure.

(3) Every disjoint family of nonzero subprojections of P_ϕ is countable.

(4) $P_\phi = P_\mu$ for some μ in $ca(\mathcal{A})^+$.

(5) ϕ has a control measure.

Proof. By 5.2, (1) and (2) are equivalent, while (2), (3), and (4) are equivalent by 5.5. By 1.4(1) and 1.5(2), (4) and (5) are equivalent.

The equivalence of (1) and (5) is due to Drewnowski [4, Theorem 2.3], who proves it using Fréchet-Nikodým topologies.

References

1. C. H. Brook, Projections and measures, thesis, University of North Carolina, Chapel Hill, N. C., 1978.

2. ________________ and W. H. Graves, The range of a vector measure, J. Math. Anal. Appl., to appear.

3. L. Drewnowski, Topological rings of sets, continuous set functions, integration. I, Bull. Acad. Polon. Sci. Sér. Math. Astronom. Phys. 20 (1972), 269-276.

4. ________________, On control submeasures and measures, Studia Math. 50 (1974), 203-224.

5. W. H. Graves, On the theory of vector measures, Amer. Math. Soc. Mem., no. 195, 1977.

6. P. R. Halmos, Lectures on Boolean Algebras, Van Nostrand, Princeton, 1963.

7. I. Kluvánek, The range of a vector-valued measure, Math. Systems Theory 7 (1973), 44-54.

8. ________________ and G. Knowles, Vector Measures and Control Systems, North-Holland, Amsterdam, 1975.

9. B. J. Pettis, Linear functionals and completely additive set functions, Duke Math. J. 4 (1938), 552-565.

Department of Mathematical Sciences
Northern Illinois University
DeKalb, Illinois 60115

Department of Mathematics
University of North Carolina
Chapel Hill, North Carolina 27514

Contemporary Mathematics
Volume 2
1980

WEAK AND STRONG COMPACTNESS IN THE SPACE OF PETTIS INTEGRABLE FUNCTIONS

by

James K. Brooks and Nicolae Dinculeanu

Introduction

In Dunford and Schwartz's book [8, Th. IV. 8.18] relatively strong compact sets K in the space $L^p(\mu)$, $1 \leq p < \infty$, have been characterized by means of uniform (with respect to $f \in K$) strong convergence $E_\pi f \to f$, of conditional expectations E_π, determined by the net of finite partitions π. Recently, strong and weak uniform convergence of conditional expectations determined by finite or countable partitions, has been used in [5] to give characterizations of relatively weakly compact or conditionally weakly compact subsets of the spaces $L^p_E(\mu)$, $1 \leq p < \infty$, of E-valued functions, where E is a Banach space. (A set K is conditionally weakly compact if every sequence from K contains a weak Cauchy subsequence. A set is relatively weakly compact if its weak closure is weakly compact). The use of conditional expectations has the advantage of dispensing with the Radon-Nikodym property of the spaces E and E' required in previous criteria of weak compactness in $L^p_E(\mu)$.

Copyright © 1980, American Mathematical Society

In this paper we shall extend the same method to the space $L_E^1(\mu)$ of strongly measurable Pettis integrable functions and shall give conditions for sets K in $L_E^1(\mu)$ to be conditionally weakly compact, by means of weak convergence $E_\pi f \to f$ in $L_E^1(\mu)$, uniformly for $f \in K$ (theorem 1). In theorems 5 and 6 we give additional sufficient conditions for a conditionally weakly compact set to be relatively weakly compact. Then (theorems 7 and 8) we apply these results to the space U_E^1 of unconditionally convergent series of elements from E and give necessary and sufficient conditions of relative weak compactness and conditional weak compactness in U_E^1. In section 5 conditionally strongly compact sets in $L_E^1(\mu)$ or U_E^1 are characterized by means of uniform strong convergence of conditional expectations. Finally, in section 6 we extend the previous results to the space of measures with finite semi-variation.

1. Preliminaries

Let (X,Σ,μ) be an arbitrary measure space, where Σ is a σ-algebra; let E be a Banach space. Denote by Σ_f the class of sets $A \in \Sigma$ with $\mu(A) < \infty$. We shall consider μ-partitions $\pi = (A_i)_{i\in I}$ with $A_i \in \Sigma_f$ and $\mu(A_i) > 0$. The union of the sets A_i is not necessarily equal to X. If $\pi' = (B_j)_{j\in J}$ is a similar partition, write $\pi \le \pi'$ if every B_j is either contained in some A_i, or disjoint from all A_i, and if $\cup A_i \subset \cup B_j$.

Let $L_E^1(\mu)$ be the space of strongly μ-measurable Pettis-integrable functions $f:X \to E$, endowed with the norm

$$(f)_1 = \sup\{\int|\langle f,x'\rangle|d\mu;\ x' \in E,\ |x'| \le 1\}.$$

Let $\pi = (A_i)_{i\in I}$ be a μ-partition and $f \in L^1_E(\mu)$. We define the conditional expectation $E_\pi f$ by

$$E_\pi f = \Sigma_{i\in I}[\mu(A_i)]^{-1}(\int_{A_i} fd\mu)\phi_{A_i}.$$

Note that the set $\{i \in I: \int_{A_i} fd\mu \ne 0\}$ is at most countable, since $\int_{(\cdot)} fd\mu$ is σ-additive. It can be shown that the following properties hold:

1.) $\langle E_\pi f,x'\rangle = E_\pi\langle f,x'\rangle$, for $x' \in E'$;

2.) $E_\pi f$ is Pettis-integrable and for $A \in \Sigma$ we have

$$\int_A E_\pi fd\mu = \Sigma_{i\in I}\mu(A \cap A_i)[\mu(A_i)]^{-1}\int_{A_i} fd\mu;$$

3.) E_π is a contraction on $L^1_E(\mu)$, that is, $(E_\pi(f))_1 \le (f)_1$;

4.) $E_\pi f \to f$ strongly in $L^1_E(\mu)$, where (π) is any net of μ-partitions cofinal to the net of all finite partitions over any ring R of sets of finite measure generating the δ-ring Σ_f. To prove this, it suffices to show that the simple functions over R are dense in $L^1_E(\mu)$. First of all, f has σ-finite support. To see this, let (x^*_n) be a weak star dense subset of the unit sphere of E, which can be assumed to be separable, since f has essentially separable range. Since each $x^*_n f \in L^1(\mu)$ has σ-finite support, it follows that f has σ-finite support. Now we use a representation theorem [3] which states that $f = g + \Sigma_i x_i\phi_{E_i}$ μ.a.e., where g is a bounded

Bochner integrable function and (E_i) is a disjoint sequence of sets such that $\Sigma x_i \mu(E_i)$ converges unconditionally. Obviously g and $\sum_i x_i \phi_{E_i}$ can be approximated in $L^1_E(\mu)$ by simple functions over R, whence our assertion follows.

In the particular case where X is the set N of natural numbers and μ is the counting measure, $L^1_E(\mu)$ is the space U^1_E of unconditionally convergent series in E. If π_k is the partition {1}, {2}, ... , {k}, then $E_{\pi_k}(x)$ is the projection $P_k(x) = P_k(x_1, x_2, \ldots) = (x_1, \ldots, x_k, 0, \ldots)$.

2. Conditional weak compactness in $L^1_E(\mu)$.

This section contains the main results of this paper, namely the characterization of conditional weak compactness by means of uniform weak convergence of conditional expectations. We shall also give a characterization of uniform weak convergence for a family of sequences in a tensor product space, which extends results in [10].

Theorem 1. Let $K \subset L^1_E(\mu)$. If:

(1) For each set $A \in \Sigma$ of finite measure, the set $K(A) = \{\int_A f d\mu; f \in K\}$ is conditionally weakly compact in E;

(2) For every countable subset $K' \subset K$, there exists a countable subset $K_0 \subset K'$ and a sequence (π_n) of finite μ-partitions such that $E_\pi f \to f$ weakly in $L^1_E(\mu)$, uniformly for $f \in K_0$;

then K is conditionally weakly compact in $L^1_E(\mu)$.

Conversely, if K is conditionally weakly compact, then condition (1) is satisfied. If, in addition, for every countable subset $K' \subset K$ there is a countable subset $K_0 \subset K'$ such that $\sup\{|f(t)|;\ f \in K_0\} < \infty$, μ-a.e., then condition (2) is also satisfied.

Proof. First assume conditions (1) and (2). If $\pi = (A_1,\ldots,A_k)$ is any finite μ-partition, then $E_\pi(K)$ is contained in the set $\Sigma_{1\le i\le k}\mu(A_i)^{-1}K(A_i)\phi_{A_i}$, which by hypothesis (1) is conditionally weakly compact. In particular, for the partitions (π_n) in hypothesis (2), we deduce that the corresponding sets $E_{\pi_n}(K)$ are conditionally weakly compact. By lemma 6 in [5], conditional weak compactness is preserved by uniform weak convergence, therefore any countable subset $K' \subset K$ contains a conditionally weakly compact subset $K_0 \subset K'$; therefore K is conditionally weakly compact.

Conversely, assume that K is conditionally weakly compact. Condition (1) follows from the continuity of the mapping $f \to \int_A f d\mu$ of $L^1_E(\mu)$ into E. Now let $K' \subset K$ be a countable subset and let $K_0 \subset K'$ be such that $\sup\{|f(t)|;\ f \in K_0\} < \infty$, μ-a.e. Then there is an increasing sequence (C_n) of sets of finite measure such that all functions $f \in K$ vanish μ-a.e. outside of the union $X_0 = \cup_n C_n$ and such that for each n, we have $\sup\{|f(t)|;\ f \in K_0\} \le n$ for all $t \in C_n$. Let Σ_0 be the separable σ-algebra of subsets of X_0 with respect to which all functions $f \in K_0$ are measurable, and let $\mathcal{R}_0$ be a countable ring of sets of finite measure, containing all sets C_n and generating the σ-ring Σ_0 (see [6]). Finally, let (π_n) be a sequence of finite μ-partitions over R_0, cofinal to the net of all finite μ-partitions over R_0 (see step J of the proof of theorem 1 in [5]). For each $x' \in E'$, with $|x'| \le 1$ and each partition π_n, the set $\langle K,x'\rangle = \{\langle f,x'\rangle; f \in K\}$ is conditionally weakly compact in $L^1(\mu)$, by the continuity of the mapping $f \to \langle f,x'\rangle$ of $L^1_E(\mu)$ into $L^1(\mu)$.

The functions $\langle f,x'\rangle$ of $\langle K_0,x'\rangle$ vanish μ-a.e. outside X_0, are measurable with respect to Σ_0, and, moreover, for each n we have $\sup\{|\langle f(t),x'\rangle|; f \in K_0\} \le n$ for all $t \in C_n$. Therefore we can use the countable ring R_0 and the sequence (π_n) to apply theorem 1 in [6] and deduce that

$$\langle E_{\pi_n} f,x'\rangle \to \langle f,x'\rangle, \text{ as } n \to \infty$$

weakly in $L^1(\mu)$, uniformly for $f \in K_0$; that is, for each $g \in L^\infty(\mu)$ we have

$$(*) \quad \int\langle E_{\pi_n} f,x'\rangle g\,d\mu \to \int\langle f,x'\rangle g\,d\mu, \text{ as } n \to \infty,$$

uniformly for $f \in K_0$. It follows then that (*) is valid for any $x' \in E'$. We shall prove now that for any continuous linear functional L on $L^1_E(\mu)$ we have

$$\langle E_{\pi_n} f,L\rangle \to \langle f,L\rangle \text{ as } n \to \infty$$

uniformly for $f \in K_0$, which means that $E_{\pi_n} f \to f$ weakly in $L^1_E(\mu)$, uniformly for $f \in K_0$.

Now let L be such a continuous linear functional on $L^1_E(\mu)$. Let B_1 be the w*-closure of a norming set in the unit ball of E' and let B_2 be the w*-closure in $L^\infty(\mu)$

of a norming set in the unit ball of $L^\infty(\mu)$, for example, the w*-closure of the step functions $\Sigma_{1\le i\le n}\phi_{A_i}\alpha_i$, with $A_i \in R$ disjoint and $|\alpha_i| \le 1$, where R is as in 4. section 1.

Consider on B_1 and B_2 the respective w*-topologies. Then B_1 and B_2 are compact. Define $B = B_1 \times B_2$; then B is compact with the product topology. Every function $f \in L_E^1(\mu)$ can be considered as a bounded function defined on B by the following equality

$$f(x',g) = \int\langle f,x'\rangle g d\mu, \quad x' \in B_1, \; g \in B_2.$$

Moreover, the norm of f in $L_E^1(\mu)$ is equal to the sup norm on f on B:

$$\begin{aligned}(f)_1 &= \sup\{\int|\langle f,x'\rangle|d\mu;\ x' \in B_1\} = \\ &= \sup\{|\int\langle f,x'\rangle g d\mu;\ x' \in B_1,\ g \in B_2\} = \\ &= \sup\{|f(x',g)|;\ (x',g) \in B\}.\end{aligned}$$

If $f = \Sigma\phi_{A_i}x_i$ is a step function from $L_E^1(\mu)$, then f is continuous on B, since $f(x',g) = \Sigma\langle x_i,x'\rangle\langle\phi_{A_i},g\rangle$, for $(x',g) \in B$, where $\langle\phi,g\rangle = \int\phi g d\mu$ for $\phi \in L^1(\mu)$ and $g \in L^\infty(\mu)$.

If now $f \in L_E^1(\mu)$, there exists a sequence (f_n) of step functions converging to f in $L_E^1(\mu)$, that is, converging to f uniformly on B. Since the f_n are continuous on B, it follows that f is also continuous on B. In this way, $L_E^1(\mu)$ can be embedded isometrically in the space C(B) of the scalar valued continuous functions on B. By the Hahn-Banach theorem, L can be extended to a continuous linear functional on C(B), still denoted by L. By the Riesz-Kakutani representation theorem, there exists a scalar valued regular Borel measure λ on B such that

$$\langle f,L\rangle = \int_B f(x',g)d\lambda(x',g), \text{ for } f \in C(B).$$

In particular, for $f \in L^1_E(\mu)$ we have

$$\langle E_{\pi_n}(f),L\rangle = \int_B (E_{\pi_n}(f))(x'g)d\lambda(x',g) =$$

$$= \int_B (\int \langle E_{\pi_n}(f),x'\rangle g d\mu)d\lambda(x',g).$$

The relation (*) means that for each $x' \in E'$ and each $g \in L^\infty(\mu)$, we have

$$(E_{\pi_n}(f))(x',g) \to f(x',g), \text{ as } n \to \infty, \text{ uniformly}$$

for f in K_0. Let $\varepsilon > 0$. The proof of Egorov's theorem can be easily adapted to prove that there exists a set $B_0 \subset B$ with $\lambda(B-B_0) < \varepsilon$ and

$$(E_{\pi_n}(f)(x',g) \to f(x',g) \text{ as } n \to \infty$$

uniformly for $(x',g) \in B_0$ and uniformly for $f \in K_0$. Let n_ε be such that

$$|(E_{\pi_n}(f))(x',g) - f(x',g)| < \varepsilon \text{ for } (x',g) \in B_0, \ f \in K_0$$

and $n \geq n_\varepsilon$. Then, for $n \geq n_\varepsilon$ and $f \in K_0$ we have

$$|\langle E_{\pi_n}(f),L\rangle - \langle f,L\rangle| = |\int_B ((E_{\pi_n})f(x',g) - f(x',g))d\lambda| \leq$$

$$\leq |\int_{B_0} ((E_{\pi_n} f)(x',g) - f(x',g))d\lambda| +$$

$$+ |\int_{B-B_0} ((E_{\pi_n} f)(x',g) - f(x',g))d\lambda| \leq$$

$$\leq \varepsilon\lambda(B_0) + 2\lambda(B-B_0)M \leq \varepsilon\lambda(B) + 2\varepsilon M,$$

where $M = \sup\{(f)_1 : f \in K\} < \infty$, since K is bounded in $L^1_E(\mu)$. From the above computation we deduce that $\langle E_{\pi_n}(f), L\rangle \to \langle f, L\rangle$ as $n \to \infty$, uniformly for $f \in K_0$. Since L was an arbitrary continuous linear functional on $L^1_E(\mu)$, it follows that $E_{\pi_n}(f) \to f$ weakly in $L^1_E(\mu)$, uniformly for $f \in K_0$, and condition (2) is satisfied.

Remark. The space $L^1_E(\mu)$ is a dense subspace of the inductive tensor product $L^1(\mu) \hat{\otimes}_\lambda E$. The proof of the last part of the above theorem remains valid for the following general situation:

Lemma 2. Let E and F be two Banach spaces, let I be any set, and for each $\alpha \in I$ let $(u_{n,\alpha})_{1 \le n < \infty}$ be a sequence from $E \hat{\otimes}_\lambda F$ such that $\sup\{||u_{n,\alpha}||; n \in N, \alpha \in I\} < \infty$. Then $u_{n,\alpha} \to 0$ as $n \to \infty$, weakly in $E \hat{\otimes}_\lambda F$, uniformly for $\alpha \in I$ if and only if for every $x' \in E'$ and every $y' \in F'$ we have $\langle u_{n,\alpha}, x' \otimes y'\rangle \to 0$ as $n \to \infty$, uniformly for $\alpha \in I$.

In fact, we can take B_1 to be the unit ball of E', B_2 the unit ball of F' with the respective w*-topologies and $B = B_1 \times B_2$. Then $E \otimes_\lambda B$ can be embedded isometrically in C(B) by identifying $u \in E \otimes_\lambda F$ with the function $(x', y') \to \langle u, x' \otimes y'\rangle$ on B. The proof follows along the same line as the preceeding proof.

Remark. This lemma extends to uniform convergence of a family of sequences the corresponding lemma proved for one sequence in [10]. In this last case, the condition $\sup\{||u_{n,\alpha}||: n \in N, \alpha \in I\} < \infty$ is superfluous.

The preceeding lemma is valid if the last condition is satisfied for $x' \in B_1$ and $y' \in B_2$, where B_1 and B_2 are weak star closed norming sets in the unit balls of E and F respectively.

In the particular case of the space $L_E^1(\mu)$ we have a slightly stronger result:

Lemma 3. Let I be any set and for each $\alpha \in I$ let $(f_{n,\alpha})_{1 \le n < \infty}$ be a sequence from $L_E^1(\mu)$ such that $\sup\{(f_{n,\alpha})_1 : n \in N, \alpha \in I\} < \infty$. Then $f_{n,\alpha} \to 0$ as $n \to \infty$, weakly in $L_E^1(\mu)$, uniformly for $\alpha \in I$, if and only if for every $A \in \Sigma$ and for every $x' \in E'$ we have $\int_A \langle f_{n,\alpha}, x' \rangle d\mu \to 0$, as $n \to \infty$, uniformly for $\alpha \in I$.

Proof. Assume the last condition is satisfied. Then for every $x' \in E'$ and every step function $g = \Sigma_{1 \le i \le n} \phi_{A_i} \alpha_i$, with $A_i \in \Sigma$ and α_i scalars, we have $\int \langle f_{n,\alpha}, x' \rangle g d\mu \to 0$ as $n \to \infty$, uniformly for $\alpha \in I$. The last relation remains valid for functions $g \in L^\infty(\mu)$ since the step functions are dense in $L^\infty(\mu)$ and since the set $\{f_{n,\alpha} : n \in N, \alpha \in I\}$ is bounded in $L_E^1(\mu)$.

3. Relative weak compactness in $L_E^1(\mu)$.

In this section we shall prove first a lemma which gives additional sufficient conditions for a conditionally weakly compact set in $L_E^1(\mu)$ to be relatively weakly compact; from this lemma we shall derive various theorems giving sufficient conditions for relative weak compactness in $L_E^1(\mu)$.

Lemma 4. A set $K \subset L^1_E(\mu)$ is relatively weakly compact if it satisfies the following conditions:

(a) K is conditionally weakly compact;

(b) For every $A \in \Sigma$, the set $K(A) = \{\int_A f d\mu : f \in K\}$ is relatively weakly compact in E;

(c) Every countable set $K' \subset K$ contains a countable subset $K_0 \subset K'$ such that $\sup\{|f(t)|; f \in K_0\} < \infty$, μ-a.e.;

(d) Any measure $m: \Sigma \to E$ of the form $m(A) = \lim_n \int_A f_n d\mu$, (weak limit in E), for some sequence (f_n) from K and for any $A \in \Sigma$, has a Radon-Nikodym derivative with respect to μ, on each set of finite m-variation.

Proof. Let (f_n) be a weak Cauchy sequence from K; we shall prove that there exists an $f \in L^1_E(\mu)$ such that $f_n \to f$ weakly in $L^1_E(\mu)$. Without loss of generality we may assume that $\sup_n |f_n(t)| < \infty$, μ-a.e. (by hypothesis (c)) and that Σ is generated by a countable ring R of sets of finite measure. In particular, we may assume that X is a countable union of disjoint sets X_k of finite measure such that $\sup_n\{|f_n(t)|: t \in X_k\} = M_k < \infty$, for each k. For every set $A \in \Sigma$, the sequence $(\int_A f_n d\mu)$ is weak Cauchy in K(A), therefore, by hypothesis (b), there exists an $m(A) \in E$ such that

$$\langle m(A), x' \rangle = \lim_n \int_A \langle f_n, x' \rangle d\mu, \text{ for } x' \in E'.$$

The set function $m: \Sigma \to E$ is additive and weakly σ-additive, by the Nikodym theorem. Then, by Pettis' theorem, m is strongly σ-additive. Moreover, m has σ-finite variation on every set X_k. In fact, let $(B_1,\ldots,B_p)$ be disjoint sets from Σ_f contained in X_k. For each $\varepsilon > 0$ and each i, there exists $x_i' \in E'$ with $|x_i'| \le 1$ and $|m(B_i)| < \langle m(B_i), x_i'\rangle + \frac{\varepsilon}{p}$. Then $\Sigma_{1\le i\le p}|m(B_i)| - \varepsilon < \Sigma_{1\le i\le p} \langle m(B_i), x_i'\rangle =$

$\Sigma_{1\le i\le p} \lim_n \int_{B_i} \langle f_n,x_i'\rangle\, d\mu \le \Sigma_{1\le i\le p} \liminf_n \int_{B_i} |\langle f_n,x_i'\rangle|\, d\mu \le$

$\le \liminf_n \Sigma_{1\le i\le p} \int_{B_i} |\langle f_n,x_i'\rangle|\, d\mu \le M_k\mu(X_k) < \infty;$

hence m has finite variation on X_k. By hypothesis (d) there exists a Bochner integrable function $g_k\colon X \to E$ vanishing outside X_k, such that

$$m(B) = \int_B g_k\, d\mu, \text{ for } B \in \Sigma \cap X_k.$$

The function $f = \Sigma_{1\le k<\infty}\, g_k$ is μ-measurable and Pettis integrable, and we have

$$m(A) = \int_A f d\mu, \text{ for } A \in \Sigma.$$

In fact, let $A \in \Sigma$ and $x' \in E'$. The scalar measure $m_{x'}$ defined on Σ by $m_{x'}(B) = \langle m(B),x'\rangle$ for $B \in \Sigma$ satisfies the following equality:

$$m_{x'}(A_k \cap A) = \langle m(A \cap A_k),x'\rangle = \int_{A\cap A_k} \langle f,x'\rangle d\mu,$$

and $m_{x'}$ has finite varation on A. From [7] lemma 2 we deduce that $\langle f,x'\rangle\phi_A$ is μ-integrable and

$$\langle m(A),x'\rangle = m_{x'}(A) = \int_A \langle f,x'\rangle d\mu.$$

Since $A \in \Sigma$ and $x' \in E'$ are arbitrary, it follows that f is Pettis integrable and

$$\int_A \langle f,x'\rangle d\mu = \lim \int_A \langle f_n,x'\rangle d\mu$$

for $A \in \Sigma$ and $x' \in E'$. By Cor. 3.4 in [10], $f_n \to f$ weakly in $L^1_E(\mu)$; hence K is relatively weakly compact.

Remarks. Conditions a, b and d are also necessary for relative weak compactness. This is evident for conditions a and b. To see that condition d holds, let $m(A) = \text{(weak)} \lim \int_A f_n d\mu = \int_A f d\mu$, where $f \in L^1_E(\mu)$. We may assume that m has finite variation on Σ. Again, by the representation theorem of f used in the proof of property 4. section 1, $f = g + \Sigma_{i=1}^{\infty} x_k \phi_{E_i}$, hence the E-valued set function

$\lambda(A) = \Sigma_{i=1}^{\infty} x_i \mu(E_i \cap A) = m(A) - \int_A g d\mu$ has finite variation, since the E_i are disjoint, that is $\Sigma_{i=1} |x_i| \mu(E_i \cap A) < \infty$. Hence f is Bochner integrable and condition (d) follows.

Combining lemma 4 and theorem 1 and Cor. 3.5 in [10], we obtain the following.

Theorem 5. Assume that E has the Radon Nikodym property. Let $K \subset L^1_E(\mu)$ be a set such that for any countable set $K' \subset K$ there is a countable subset $K_0 \subset K'$ with $\sup \{|f(t)|; f \in K_0\} < \infty$, μ-a.e. Then K is relatively weakly compact in $L^1_E(\mu)$ if and only if K satisfies condition 1 and either one of the conditions 2 or 2' below:

1) For every $A \in \Sigma$, the set $K(A) = \{\int_A f d\mu;\ f \in K\}$ is relatively weakly compact in E;

2) For every countable set $K' \subset K$, there exists a countable subset $K_0 \subset K'$ and a sequence (π_n) of finite partitions such that $E_{\pi_n} f \to f$ weakly in $L^1_E(\mu)$, uniformly for $f \in K_0$.

2') For every $x' \in E'$, the set $\langle K,x' \rangle = \{\langle f,x' \rangle; f \in K\}$ is relatively weakly compact in $L^1(\mu)$.

We can relax the restriction on E to have the Radon-Nikodym

property, if we strengthen condition 1:

Theorem 6. Assume that $\mu(X) < \infty$. Let $K \subset L^1_E(\mu)$ be a set such that for any countable set $K' \subset K$, there exists a countable subset $K_0 \subset K'$ satisfying $\sup\{|f(t)|; f \in K_0\} < \infty$, μ-a.e. Then K is relatively weakly compact in $L^1_E(\mu)$ if it satisfies condition 1 and either one of the conditions 2 and 2' below:

1) The set $H = \{[\mu(A)]^{-1} \int_A f d\mu;\ f \in K,\ A \in \Sigma,\ \mu(A) > 0\}$ is relatively weakly compact in E;

2) For every countable set $K' \subset K$ there exists a countable subset $K_0 \subset K'$ and a sequence (π_n) of finite partitions such that $E_{\pi_n} f \to f$ weakly in $L^1_E(\mu)$, uniformly for $f \in K_0$;

2') For every $x' \in E'$, the set $\langle K, x'\rangle = \{\langle f, x'\rangle;\ f \in K\}$ is relatively weakly compact in $L^1(\mu)$.

Proof. We remark that for every $A \in \Sigma$, we have $K(A) \subset \mu(A)H$, therefore $K(A)$ is relatively weakly compact in E. From theorem 1 and from [10] it follows that K is conditionally weakly compact, so that hypothesis a and b of lemma 5 are satisfied. The assumption in the present theorem implies also hypothesis c in lemma 4. In order to prove that hypothesis d in lemma 4 is satisfied, let $m\colon \Sigma \to E$ be a measure of the form $\langle m(A), x'\rangle = \lim_n \int_A \langle f_n, x'\rangle d\mu$, for $A \in \Sigma$, $x' \in E'$, with $f_n \in K$. Then, for each $A \in \Sigma$, the quotient $m(A)/\mu(A)$ belongs to the weak closure in E of the set H, which is weakly compact. It follows that the average range $\{m(A)/\mu(A);\ A \in \Sigma, \mu(A) > 0\}$ is relatively weakly compact. By the Metivier [11] version of the Radon-Nikodym theorem, m has Radon-Nikodym derivative on every set of finite m-variation. We can now apply lemma 4 and deduce that K is relatively weakly compact.

Remark. Unlike condition 1 in theorem 6, condition 1 in theorem 7 is, in general, not necessary for K to be relatively weakly compact.

4. Weak compactness in the space U_E^1

In this section we shall study conditional weak compactness and relative weak compactness in the Banach space U_E^1 of unconditionally convergent series (x_i) of elements from E, with norm $\|(x_i)\|_1 = \sup\{\Sigma_{i=1}^{\infty}|\langle x_i, x'\rangle|;\ |x'| \le 1\}$. We begin with the characterization of conditionally weakly compact sets.

Theorem 7. *A set* $K \subset U_E^1$ *is conditionally weakly compact if and only if:*

1) *For each* $i \in N$, *the set* $K(i) = \{x_i;\ x = (x_i) \in K\}$ *is conditionally weakly compact in* E;

2) *For each* $x' \in E'$, *the set* $\langle K, x'\rangle = \{(\langle x_i, x'\rangle)_{1 \le i < \infty};\ x = (x_i) \in K\}$ *is relatively weakly compact in* ℓ^1.

Proof. The theorem follows from Cor. 3.6 in [10], but we shall give a direct proof.

Assume first that conditions 1 and 2 are satisfied. Then $P_k x \to x$ as $k \to \infty$, weakly, uniformly for $x \in K$, according to condition 2 (see for example theorem 2 in [4]). Also $P_k K \subset \Sigma_{i \le k} K(i)\delta_i$, where δ_i is the sequence having 1 at the ith place and zero elsewhere. Since the sets K(i) are conditionally weakly compact E, the sets $K(i)\delta_i$ are conditionally weakly compact in U_E^1, therefore $P_k K$ is conditionally weakly compact in U_E^1. We can then apply lemma 6 in [5] and deduce that K is conditionally weakly compact in U_E^1. Conversely, assume that K is conditionally weakly compact. Then condition 1 is evidently satisfied, and condition 2 follows from the continuity of the mapping $(x_i) \to (\langle x_i, x'\rangle)$ of U_E^1 into ℓ^1.

Remark. Condition 2 is equivalent to each one of the following two conditions:

2') For each $x' \in E'$ we have $\Sigma_{i\geq k}|\langle x_i, x'\rangle| \to 0$ as $k \to \infty$, uniformly for $x = (x_i) \in K$;

2") For each $x' \in E$; and each $a = (a_i) \in \ell^\infty$ we have $\Sigma_{i\geq k}\langle x_i, x'\rangle a_i \to 0$ as $k \to \infty$, uniformly for $x = (x_i) \in K$.

The following theorem gives a characterization of relatively weakly compact sets in U_E^1:

Theorem 8. A set $K \subset U_E^1$ is relatively weakly compact if and only if it satisfies condition 2 and any one of the conditions 1, 1', 1" below:

1) For every $A \subset N$, the set $K(A) = \{\Sigma_{i\in A} x_i;\ x = (x_i) \in K\}$ is relatively weakly compact in E;

1') The set $\cup_{n\geq 1}\{\Sigma_{i\leq n} x_i;\ (x_i) \in K\}$ is relatively weakly compact in E;

1") The set $\{\Sigma_{1\leq i<\infty} a_i x_i;\ x = (x_i) \in K,\ a = (a_i) \in \ell^\infty,\ ||a||_\infty \leq 1\}$ is relatively weakly compact in E;

2) For every $x' \in E$;, the set $\langle K, x'\rangle = \{(\langle x_i, x'\rangle)_{1\leq i<\infty},\ x = (x_i) \in K\}$ is relatively weakly compact in ℓ^1.

Proof. We remark first that each one of the conditions 1, 1' and 1" implies that for each $i \in N$, the set $K(i)$ is relatively weakly compact. From theorem 7 we deduce that any one of the conditions 1, 1', or 1" together with condition 2 imply that K is conditionally weakly compact. We also remark that condition 1" implies conditions 1 and 1'.

Assume first conditions 1 and 2 and prove that K is relatively weakly compact. Let (x^n) be a sequence from K; since K is conditionally weakly compact, we may assume that (x^n) is weak Cauchy. Then, for every $i \in N$, the sequence $(x_i^n)_{1 \leq n < \infty}$ is weak Cauchy in $K(i)$, which is relatively weakly compact; hence there is $x_i^0 \in E$ such that $x_i^n \to x_i^0$ as $n \to \infty$, weakly in E. The sequence $x^0 = (x_i^0)_{1 \leq i < \infty}$ belongs to U_E^1. In fact, for any set $A \in \Sigma$ (the σ-algebra of all subsets of N), the sequence $m_n(A) = \Sigma_{i \in A} x_i^n$ is weakly Cauchy in $K(A)$, which is relatively weakly compact; therefore there exists $m(A) \in E$ such that $m_n(A) \to m(A)$ as $n \to \infty$, weakly in E, that is

$$<m_n(A), x'> \to <m(A), x'> \text{ as } n \to \infty,$$

for $A \in \Sigma$ and $x' \in E'$. From the Nikodym theorem, it follows that $<m(.), x'>$ is σ-additive; by the Pettis theorem we deduce that m is strongly σ-additive. In particular

$$m(A) = \Sigma_{i \in A} m(i) = \Sigma_{i \in A} x_i^0 \in E.$$

Therefore $x^0 \in U_E^1$. From the equality

$$\lim_n \Sigma_{i \in A} <x_i^n, x'> = <m(A), x'> = \Sigma_{i \in A} <x_i^0, x'>$$

for any $A \in \Sigma$ and $x' \in E'$ and from lemma 3 we deduce that

$x^n \to x^0$ weakly in U^1_E; therefore K is relatively weakly compact.

Assume now that conditions 1' and 2 are satisfied. Let again (x^n) be a weak Cauchy sequence from K and prove that it converges weakly to some element $x^0 \in U^1_E$. For each $i \in N$, the sequence $(x^n_i)_{1 \le n < \infty}$ is weak Cauchy in K(i), which is relatively weakly compact; therefore, there exists $x^0_i \in E$ such that

$$(1) \quad \lim_n \langle x^n_i, x'\rangle = \langle x^0_i, x'\rangle, \text{ for } x' \in E'.$$

For each $k \in N$ set $z_k = \Sigma_{1 \le i \le k} x^k_i$. The sequence $(z_k)_{1 \le k < \infty}$ belongs to a set which is relatively weakly compact in E; therefore there exists an element $z \in E$ and a subsequence $(z_{n_j})_{1 \le j < \infty}$ such that $z_{n_j} \to z$ as $j \to \infty$, weakly in E.

We prove now that for every $x' \in E'$ we have

$$(2) \quad \lim_j \Sigma_{1 \le i < \infty} \langle x^{n_j}_i, x'\rangle = \lim_j \Sigma_{1 \le i \le n_j} \langle x^{n_j}_i, x'\rangle =$$

$$= \lim_j \langle z_{n_j}, x'\rangle = \langle z, x'\rangle.$$

Let $x' \in E'$ and $\varepsilon > 0$; since $\{(\langle x^n_i, x'\rangle)_{1 \le i < \infty}\}_{1 \le n < \infty}$ is a weak Cauchy sequence in ℓ^1, there exists k_ε such that

$$(3) \quad \Sigma_{i \ge k_\varepsilon} |\langle x^n_i, x'\rangle| \le \varepsilon, \text{ for all } n.$$

Let j_ε be such that $n_{j_\varepsilon} > k_\varepsilon$. Then, if $j > j_\varepsilon$, we have $n_j \ge k_\varepsilon$, hence

$$\Sigma_{i \ge n_j} |\langle x^{n_j}_i, x'\rangle| \le \varepsilon,$$

therefore

$$|\Sigma_{1\leq i<\infty} < x_i^{n_j}, x'>-\Sigma_{1\leq i<n_j} < x_i^{n_j}, x'>| =$$

$$= |\Sigma_{i\geq n_j} < x_i^{n_j}, x'| \leq \varepsilon$$

from which (2) above follows.

Since from (3) it follows that the series $\Sigma_{1\leq i<\infty} < x_i^n, x'>$ converges uniformly with respect to n, from (1) and (2) above we deduce

$$\Sigma_{1\leq i<\infty} <x_i^0, x'> = \Sigma_{1\leq i<\infty} \lim_n <x_i^n, x'> =$$

$$= \lim_n \Sigma_{1\leq i<\infty} <x_i^n, x'> = \lim_j \Sigma_{1\leq i<\infty} <x_i^{n_j}, x'> = <z,x>.$$

Since the series $\Sigma_{1\leq i<\infty} x_i^0$ is weakly unconditionally convergent, it is unconditionally convergent, by the Orlicz-Pettis theorem; therefore $x^0 = (x_i^0) \in U_E^1$.

Finally, we prove that $x^n \to x^0$ weakly in U_E^1. Let $x' \in E'$ and $a = (a_i) \in \ell^\infty$ and set $M = ||a||_\infty$. Let $\varepsilon > 0$ and take k_ε such that condition (3) is satisfied:

$$\Sigma_{i\geq k_\varepsilon} |<x_i^n, x'>| \leq \varepsilon \text{ for all } n$$

and also

$$\Sigma_{i\geq k_\varepsilon} |<x_i^0, x'>| \leq \varepsilon.$$

Using (1) we can get n_ε such that for $i = 1,2,\ldots,k_\varepsilon$ and for $n \geq n_\varepsilon$ we have

$$|<x_i^n - x_i^0, x'>| < \varepsilon/k_\varepsilon,$$

hence

$$\Sigma_{i\leq k_\varepsilon} |<x_i^n - x_i^0, x'>| < \varepsilon.$$

Then, for $n \geq n_\varepsilon$ we have

$$|\Sigma_{1\leq i<\infty} < x_i^n - x_i^0, x'>a_i| \leq M\Sigma_{1\leq i<\infty}|<x_i^n - x_i^0, x'>| \leq$$

$$\leq M\{\Sigma_{i\geq k_\varepsilon}|<x_i^n - x_i^0, x'>| + \Sigma_{i\geq k_\varepsilon}|<x_i^n, x'>| +$$

$$+ \Sigma_{i\geq k_\varepsilon}|<x_i^0, x'>|\} \leq 3M\varepsilon,$$

and this means that $x^n \to x^0$ weakly in U_E^1; therefore K is relatively weakly compact.

Finally, conditions 1" and 2 also imply that K is relatively weakly compact, since condition 1" implies either of the conditions 1 and 1'.

Conversely, assume that K is relatively weakly compact. Then conditions 1 and 2 follows from the continuity of the mappings $x \to \Sigma_{i\in A}x_i$ and $x \to <x,x'>$ from U_E^1 into E or ℓ^1 respectively. Condition 1" has been proved in [10]. Therefore condition 1' also follows.

Remark. There is a gap in the proof of Corollary 3.6 in [10]; conditions (1) and (2) of that corollary ensure conditional weak compactness, but since U_E^1 is not necessarily weakly sequentially complete, an argument is needed to obtain relative weak compactness; this is carried out by means of our construction of (z_K).

5. Conditional strong compactness in $L_E^1(\mu)$ and U_E^1

We have first the following relative compactness criterion in the Banach space U_E^1:

Theorem 9. A set $K \subset U_E^1$ is relatively strongly compact if and only if:

1) For every $i \in N$, the set $K(i) = \{x_i;\ x = (x_i) \in K\}$ is relatively strongly compact in E;

2) $\Sigma_{i\geq k}|\langle x_i, x'\rangle| \to 0$ as $k \to \infty$, uniformly for $x \in K$ and $|x'| \leq 1$.

Proof. Since each P_k is a contraction and since $\lim_k P_k x = x$ for $x \in U_E^1$, we can apply Phillips' lemma [8, th. IV. 8.18] and notice that condition 2 means $P_k x \to x$ strongly in U_E^1, uniformly for $x \in K$.

Using this result, we can prove now the following conditional compactness criterion in $L_E^1(\mu)$.

Theorem 10. A set $K \subset L_E^1(\mu)$ is conditionally strongly compact if and only if:

1) For every $A \in \Sigma_f$, the set $K(A) = \{\int_A f d\mu;\ f \in K\}$ is relatively strongly compact in E;

2) There exists a net $P = (\pi)$ of μ-partitions such that $E_\pi f \to f$ strongly in $L_E^1(\mu)$, uniformly for $f \in K$;

3) The set of measures $\{\int_{(\cdot)} f d\mu;\ f \in K\}$ is uniformly σ-additive on Σ.

If P consists of finite partitions, then condition 3 is superfluous in the proof of sufficiency.

Proof. Assume first conditions 1, 2 and 3. We shall prove first that for each partition $\pi \in P$, the set $E_\pi K$ is conditionally strongly compact in $L_E^1(\mu)$.

Let $\pi = (A_i)_{i\in I} \in P$. Since we can assume that K is countable, and since $\int_{A_i} f d\mu = 0$, except for countably many indices $i \in I$, for each $f \in K$, we can also assume that the partition π is at most countable. If π is finite, $\pi = \{A_1,\ldots,A_n\}$, then $E_\pi K \subset \Sigma_{1\le i\le n}[\mu(A_i)]^{-1}K(A_i)\phi_{A_i}$, therefore $E_\pi K$ is conditionally strongly compact using condition 1, but without imposing condition 3. If π is countable, $\pi = (A_i)_{i\in N}$, we consider the linear mapping θ defined by

$$\theta E_\pi f = \left(\int_{A_i} f d\mu\right)_{1\le i<\infty}, \text{ for } f \in L^1_E(\mu).$$

First we have $\theta E_\pi f \in U^1_E$, since $\int_{(.)} f d\mu$ is σ- adiitive on Σ. Then θ is an isometry: for $x' \in E'$ we have

$$\Sigma_i |<(\theta E_\pi f)(i), x'>| = \Sigma_i |\int_{A_i} <f,x'>d\mu| =$$

$$= \int |(\Sigma_i [\mu(A_i)]^{-1}\int_{A_i} <f,x'>d\mu|\phi_{A_i})d\mu =$$

$$= \int |\Sigma_i [\mu(A_i)]^{-1}\int_{A_i} <f,x'>d\mu\phi_{A_i}|d\mu =$$

$$= \int |<E_\pi f,x'>|d\mu,$$

hence $(\theta E_\pi f)_1 = (E_\pi f)_1$. This isometry is onto U^1_E, since if $x = (x_i) \in U^1_E$, we chose $f = \Sigma_i \phi_{A_i} x_i$; then $f \in L^1_E(\mu)$, $E_\pi f = f$ and $\theta E_\pi f = x$.

The set $\theta E_\pi K$ satisfies the conditions of theorem 9 : Condition 1 of theorem 9 follows from condition 1 of the theorem since $(\theta E_\pi K)(j) = \{\int_{A_i} f d\mu;\ f \in K\}$. Then, for $f \in L^1_E(\mu)$ and $x' \in E'$ we have

$$\Sigma_{i\ge k}|<(\theta E_\pi f)(i), x'>| = \Sigma_{i\ge k}|<\int_{A_i} f d\mu, x'>| =$$

$$= \Sigma_{i\ge k}|\int_{A_i} <f,x'>d\mu \le \Sigma_{i\ge k}\int_{A_i} |<f,x'>|d\mu \to 0$$

as $k \to \infty$,

uniformly for $f \in K$ and $|x'| \leq 1$, since, by condition 3 of this theorem, the scalar measures $\{ \int_{(.)} \langle f,x'\rangle d\mu;$ $f \in K, |x'| \leq 1\}$ are uniformly σ-additive, which is equivalent to the set of positive measures $\{ \int_{(.)} |\langle f,x'\rangle| d\mu;$ $f \in K, |x'| \leq 1\}$ being uniformly σ-additive. In this way, conditions 2 of theorem 9 is also satisfied. It follows that $\theta E_\pi K$ is relatively strongly compact in U^1_E. Since θ is an isometry, we deduce that $E_\pi K$ is also relatively strongly compact in $E_\pi L^1_E(\mu) \subset L^1_E(\mu)$. Now we can apply Phillips' lemma to the set $(E_\pi)_{\pi \in P}$ of operators extended to the strong completion L of $L^1_E(\mu)$ and deduce that K is relatively strongly compact in L, hence K is conditionally strongly compact in $L^1_E(\mu)$.

Conversely, assume that K is conditionally strongly compact. Condition 1 follows from the continuity of the mapping $f \to \int_A f d\mu$ of $L^1_E(\mu)$ into E. Condition 2 follows from Phillips' lemma applied to any net $P = (\pi)$ such that $E_\pi f \to f$ strongly in $L^1_E(\mu)$, for any $f \in L^1_E(\mu)$; in particular, for any net cofinal to the net of all finite partitions over a ring generating Σ. To prove condition 3, let $(A_i)_{i \in N}$ be a sequence of disjoint sets of Σ. Consider the partition $\pi = (A_i)_{i \in N}$ and mapping ρ of $L^1_E(\mu)$ into U^1_E defined by

$$\rho f = (\int_{A_i} f d\mu)_{i \in N}, \text{ for } f \in L^1_E(\mu).$$

ρ is continuous, since for $x' \in E'$ we have

$$\Sigma_i |\langle (\rho f)(i), x'\rangle| = \Sigma_i |\langle \int_{A_i} f d\mu, x'\rangle| \leq$$

$$\leq \Sigma_i \int_{A_i} |\langle f,x'\rangle| d\mu \leq \int |\langle f,x'\rangle| d\mu$$

hence $(\rho f)_1 \leq (f)_1$. Since K is conditionally strongly compact, ρK is also conditionally strongly compact. By theorem 9, we have

$$\Sigma_{i\geq k} \left| \int_{A_i} \langle f, x' \rangle d\mu \right| \to 0, \text{ as } k \to \infty,$$

uniformly for $f \in K$ and $|x'| \leq 1$, which is equivalent to

$$\Sigma_{i\geq k} \int_{A_i} \langle f, x' \rangle d\mu \to 0, \text{ as } k \to \infty$$

uniformly for $f \in K$ and $|x'| \leq 1$, which is also equivalent to

$$\Sigma_{i\geq k} \int_{A_i} f d\mu \to 0 \text{ as } k \to \infty, \text{ uniformly for } f \in K,$$

and this means that the measures $\{ \int_{(\cdot)} f d\mu;\ f \in K\}$ are uniformly σ-additive; so, condition 3 is also satisfied.

6. Strong and weak compactness in the space of measures with norm compact range.

In this last section we extend the results of previous sections to the space $M(\Sigma,E)$ of measures $m\colon \Sigma \to E$ with finite semi-variation, endowed with the total semi-variation norm

$$\begin{aligned} ||m|| &= \sup \{||x'm||;\ x' \in E',\ |x'| \leq 1\} = \\ &= \sup \{|x'm|(X);\ x' \in E',\ |x'| \leq 1\}, \end{aligned}$$

where $|x'm|$ is the variation of $x'm$.

Let $CM(\Sigma,E)$ be the subspace of measures with norm compact range; this is the closure of the set of simple measures of the form $\Sigma_{1\leq i\leq n} x_i \mu_i$ with $x_i \in E$ and μ_i scalar measures on Σ (see [10]). Let μ be a positive measure on Σ and $m \in M(\Sigma,E)$ a measure absolutely continuous with respect to μ. For any μ-partition $\pi = (A_i)_{i\in I}$ we define

$$(E_\pi m)(A) = \Sigma_{i\in I} m(A_i)\mu(A_i)^{-1}\mu(A\cap A_i), \text{ for } A \in \Sigma.$$

If $m \in CM(\Sigma,E)$ then $E_\pi m \in CM(\Sigma,E)$. Moreover, $E_\pi m \to m$ in the norm topology of $CM(\Sigma,E)$.

Theorem 11. A set $K \subset CM(\Sigma,E)$ is relatively strongly compact if and only if:

1) For every set $A \in \Sigma$, the set $K(A) = \{m(A); m \in K\}$ is relatively strongly compact in E.

2) There exists a measure $\mu \geq 0$ on Σ such that $m << \mu$ for every $m \in K$, and a net $P = (\pi)$ of μ-partitions such that $E_\pi m \to m$ strongly in $CM(\Sigma,E)$, uniformly for $m \in K$;

3) K consists of uniformly σ-additive measures.

If, in condition 2, P consists of finite partitions, then condition 3 is superfluous in the proof of the necessity implication.

The proof is similar to that of theorem 10. We have only to remark that $CM(\Sigma,E)$ is a Banach space, hence relative strong compactness is equivalent to conditional strong compactness.

For the case of finite partitions see [4], theorem 5.2.

Theorem 12. A set $K \subset CM(\Sigma,E)$ is conditionally weakly compact if it satisfies condition 1 and either one of the conditions 2 and 2' below.

1) For every $A \in \Sigma$, the set $K(A) = \{m(A); m \in K\}$ is conditionally weakly compact in E;

2) For every countable subset $K' \subset K$ there is a countable subset $K_0 \subset K$, a positive measure μ such that $m << \mu$ for all $m \in K_0$, and a sequence (π_n) of finite partitions such that

$E_{\pi_n} m \to m$ weakly in $CM(\Sigma, E)$, uniformly for $m \in K_0$.

2') For every $x' \in E'$, the set $\langle K, x' \rangle = \{x'm; m \in K\}$ is conditionally weakly compact in the space $C(M)$ of scalar measures.

We use the fact that if K is conditionally weakly compact and E is separable, there is a measure $\mu \geq 0$ with $K \ll \mu$ (see [4], theorem 6.4).

BIBLIOGRAPHY

1. R. G. Bartle, N. Dunford and J. Schwartz, Weak Compactness and vector measures, Canadian J. Math. 7(1955), 289-305.

2. J. K. Brooks, On compactness of measures, Bull. Acad. Pol. Sci., Ser. Sci. Math. Astron. Phys., 20 (1972), 991-994.

3. ___________, Representations of weak and strong integrals in Banach spaces, Proc. Nat. Acad. Sci. (U.S.A.), 63 (1969), 266-270.

4. J. K. Brooks and N. Dinculeanu, Strong additivity, absolute continuity and compactness in spaces of vector measures, J. Math. Anal. Appl. 45 (1974), 156-175.

5. ___________, Weak compactness in spaces of Bochner integrable functions and applications, Advances in Math. 24 (1977), 172-188.

6. ___________, On weak compactness in spaces of Bochner integrable functions, Advances in Math. (to appear).

7. N. Dinculeanu and J. Uhl, A unifying Radon-Nikodym theorem for vector measures, J. Multivariate Analysis 3(1973), 184-203.

8. N. Dunford and J. Schwartz, Linear operators, Part I, Interscience, New York, 1958.

9. A. Grothendieck, Sur les applications linéaires faiblement compactes d'espaces du type C(K), Canadian J. Math. 5 (1953), 129-173.

10. D. Lewis, Conditional weak compactness in certain injective tensor products, Math. Ann. 201 (1973) 201-209.

11. M. Métivier, Martingales à valeurs vectorielles, application à la dérivation des mesures vectorielles, Ann. Inst. Fourier (Grenoble) 17 (1967), 175-208.

Department of Mathematics
University of Florida
Gainesville, Florida 32611

Contemporary Mathematics
Volume 2
1980

COMPACTNESS IN SPACES OF VECTOR-VALUED MEASURES AND A NATURAL MACKEY TOPOLOGY FOR SPACES OF BOUNDED MEASURABLE FUNCTIONS

by

William H. Graves
Department of Mathematics
University of North Carolina
Chapel Hill, North Carolina 27514

and

Wolfgang Ruess
Fachbereich 6 Mathematik
Universität Essen
Universitätstrasse 3
4300 Essen 1
West Germany

Abstract

For a locally convex space X and a subset K of the space of X-valued measures on a σ-algebra Σ, it is proved that K is relatively compact in the topology of pointwise convergence (on each $A \in \Sigma$) iff $K(A)$ is relatively compact for each $A \in \Sigma$ and K is uniformly countably additive. For a measure Φ on Σ with values in a Banach space, a recipe given for a (very unbounded) Mackey neighborhood U of 0 in the space of bounded Σ-measurable scalar-valued functions, paired with the space of scalar-valued measures on Σ, such that the set of Φ-integrals of functions in U is relatively weakly compact. Both results are tied to a description of the universal measure space of the first author as a direct limit of Buck's (ℓ^∞, β), and the second result depends on a refinement due to the second author of a theorem of Grothendieck.

Copyright © 1980, American Mathematical Society

1. Introduction

Throughout, X is a locally convex (Hausdorff) topological vector space, X^* is the continuous dual of X, $L(X,Y)$ is the space of continuous linear maps from X to Y, $L_s(X,Y)$ is $L(X,Y)$ with the topology of simple convergence (the strong operator topology), Σ is an infinite σ-algebra of subsets of a set Ω, $ca(\Sigma,X)$ is the space of all X-valued countably additive maps (measures on Σ), $ca(\Sigma)$ is $ca(\Sigma,X)$ when X is the scalar field of (real or complex) numbers, $S(\Sigma)$ is the space of scalar-valued Σ-simple functions, $(S(\Sigma),\tau)$ is $S(\Sigma)$ endowed with the (locally convex) universal measure topology τ of [8], $L(\Sigma)$ is the completion of $(S(\Sigma),\tau)$, and $B(\Sigma)$ is the space of bounded Σ-measurable scalar-valued functions on Ω.

The universal measure topology τ on $S(\Sigma)$ and $L(\Sigma)$ is studied in several papers of the first author and others including [8] where the characteristic map χ defined on Σ by $\chi(A) = \chi_A$, the characteristic function of A, is proved to be a τ-measure (with values in $S(\Sigma)$ or $L(\Sigma)$) which is universal:

(1) [8; 1.4] for every X, $ca(\Sigma,X) \simeq L((S(\Sigma),\tau),X)$ via $\Phi \longrightarrow \tilde{\Phi}$ where $\tilde{\Phi}$ is the integration map corresponding to $\Phi \in ca(\Sigma,X)$ and $\Phi = \tilde{\Phi} \circ \chi$; and

(2) [8; 1.5, 11.7] for X complete in its Mackey topology, $ca(\Sigma,X) \simeq L(L(\Sigma),X)$ via $\Phi \longrightarrow \tilde{\Phi}$ where $\tilde{\Phi}$ also denotes the extension to $L(\Sigma)$ of the integration map defined in (1). For $K \subseteq ca(\Sigma,X)$, $\tilde{K}$ is always the corresponding subset of $L((S(\Sigma),\tau),X)$ and, when X is Mackey complete, $L(L(\Sigma),X)$.

A classical theorem of Bartle, Dunford, and Schwartz from [1] asserts that $K \subseteq ca(\Sigma)$ is relatively weakly compact iff K is uniformly countably additive and $K(\Sigma)$ is bounded. It is the point of view of [8; sec. 11] that the latter conditions, which are equivalent to equicontinuity of $\tilde{K} \subseteq (S(\Sigma),\tau)^* \simeq L(\Sigma)^*$, are naturally equivalent via the Schur lemma to relative weak *-compactness of K in both weak * topologies $\sigma(ca(\Sigma),S(\Sigma))$ and $\sigma(ca(\Sigma),L(\Sigma))$ but are equivalent to weak compactness of K only because Radon-Nikodym considerations yield the deep result that $L(\Sigma)$ is semi-reflexive so that $ca(\Sigma)^* \simeq L(\Sigma)$ and the weak topology is the weak * topology $\sigma(ca(\Sigma),L(\Sigma))$. Indeed, when more generally studying weak compactness in $L^1(\mu,X)$ where X is a Banach space and $\mu \in ca(\Sigma)$ is non-negative, Radon-Nikodym considerations play a pervasive role (see, for example, [2], [3], [6]). Such studies are succinctly exposed and completely referenced in [7; Ch. IV]. The main purpose of this paper is to parlay the point of view of [8] to a solution of the problem of characterizing relative compactness in $ca(\Sigma,X)$ in the topology s of simple convergence (convergence on each $A \in \Sigma$). In section 3, it is proved for any X that $K \subseteq ca(\Sigma,X)$ is relatively s-compact iff K is uniformly countably additive and $K(A) = \{\Phi(A) : \Phi \in K\}$ is relatively compact for each $A \in \Sigma$ iff

$\tilde{K}$ is equicontinuous in $L((S(\Sigma),\tau),X)$ and $K(A)$ is relatively compact for each $A \in \Sigma$ iff $\tilde{K}$ is relatively compact in $L_s((S(\Sigma),\tau),X)$ and, when X is Mackey-complete, $L_s(L(\Sigma),X)$. From this follows a Nikodym-boundedness theorem and a Vitali-Hahn-Saks theorem.

The reader with an aversion to universal mapping properties should proceed directly to prove for himself Corollary 4 of section 2 which asserts that $ca(\Sigma,X)$ is a closed subspace, in the s topology, of a product of spaces $ca(P(\mathbb{N}),X)$ where $P(\mathbb{N})$ is the power set of the set $\mathbb{N}$ of all positive integers and then proceed to section 3. However, the reader interested in the Mackey topology $\tau(B(\Sigma),ca(\Sigma))$ should read section 2, for $S(\Sigma) \subseteq B(\Sigma) \subseteq L(\Sigma)$ by [8, 4.5 and 10.5], τ is the Mackey topology $\tau(S(\Sigma),ca(\Sigma))$ on $S(\Sigma)$ and is the Mackey topology $\tau(L(\Sigma),ca(\Sigma))$ on $L(\Sigma)$ by [8; 11.7], and it is shown in section 2 that $L(\Sigma)$ is a direct limit of spaces (ℓ^∞,β) where β is Buck's strict topology of [4] (that is, $\beta = \tau(\ell^\infty,\ell^1)$). This result is implicit in [8] and especially [16] and is at the base of the description of $\tau(B(\Sigma),ca(\Sigma)))$ given in section 4. A theorem of Diestel from [5] identifies the Banach space-valued strongly bounded maps on Σ with the weakly compact operators on $B(\Sigma)$, and a theorem of Grothendieck [9] together with results in [12] and [14] identify the countably additive among these with those operators which map a $\tau(B(\Sigma),ca(\Sigma))$-neighborhood of 0 to a relatively weakly compact set. An explanation and a recipe for this neighborhood in terms of the measure in question is given in section 4.

It is known [8; 8.3] that $L(\Sigma)$ is a projective limit of Mackey spaces $(L^\infty(\mu),\tau(L^\infty(\mu),L^1(\mu)))$ for non-negative $\mu \in ca(\Sigma)$, which lends curiosity to the description in section 2 of $L(\Sigma)$ as a direct limit of spaces (ℓ^∞,β).

2. $L(\Sigma)$ and $(S(\Sigma),\tau)$ as direct limits

Let $\mathcal{D}$ be a cofinal family of partitions of Σ into pairwise disjoint sequences of measurable sets. So $(A_i) \in \mathcal{D}$ if $\Omega = \cup A_i$, $A_i \in \Sigma$, and $A_i \cap A_j = \emptyset$ for $i \neq j$. $\mathcal{D}$ is directed by refinement: $(A_i) \leq (B_i)$ in $\mathcal{D}$ if each B_i is a subset of some A_j. For each $(A_i) \in \mathcal{D}$ define a measure (an unconditionally convergent series) $\Phi_{(A_i)}$ on $P(\mathbb{N})$ with values in $(S(\Sigma),\tau) \subseteq L(\Sigma)$ by setting $\Phi_{(A_i)}(\{n\}) = \chi_{A_n}$. Then for any subset a of $\mathbb{N}$, $\Phi_{(A_i)}(a) = \chi_A$ where A is the union of those A_n for which $n \in a$. For $(A_i) \leq (B_i)$ in $\mathcal{D}$, define a measure $\Phi_{(A_i)(B_i)}$ on $P(\mathbb{N})$ to $(S(P(\mathbb{N})),\tau) \subseteq L(P(\mathbb{N}))$ by setting $\Phi_{(A_i)(B_i)}(\{n\}) = \chi_a$ where $a = \{i : B_i \subseteq A_n\}$. In spite of the somewhat unwieldy notation, it is easy to verify that $\{\tilde{\Phi}_{(A_i)(B_i)} : (A_i) \leq (B_i) \text{ in } \mathcal{D}\}$ is a direct system of continuous linear maps and that $\tilde{\Phi}_{(A_i)} = \tilde{\Phi}_{(B_i)} \circ \tilde{\Phi}_{(A_i)(B_i)}$ for $(A_i) \leq (B_i)$ in $\mathcal{D}$. There are, in fact, two direct systems here, one tied to the spaces $(S(P(\mathbb{N})),\tau)$ and $(S(\Sigma),\tau)$ and the other to their completions $L(P(\mathbb{N}))$ and $L(\Sigma)$. Notation can be slightly simplified since according to [8; 5.1 and11.8] $L(P(\mathbb{N})) \simeq (\ell^\infty,\tau)$ as locally convex spaces where $\tau = \tau(\ell^\infty,\ell^1)$ is the Mackey topology of the $\ell^\infty - \ell^1$ pairing (Buck's strict topology [4]). Similarly, if m_0 is the space of scalar-valued sequences having finitely many values, then $(S(P(\mathbb{N})),\tau) = (m_0,\tau)$ where $\tau = \tau(m_0,\ell^1)$.

Theorem 1. The family $\{\tilde{\Phi}_{(A_i)} : (\ell^\infty,\tau) \to L(\Sigma)\}_{(A_i)\in\mathcal{D}}$ is the direct limit in the category of complete locally convex spaces of the direct system $\{\tilde{\Phi}_{(A_i)(B_i)}\}$, and in the category of locally convex spaces the direct system $\{\tilde{\Phi}_{(A_i)(B_i)}\}$ has direct limit $\{\tilde{\Phi}_{(A_i)} : (m_0,\tau) \to (S(\Sigma),\tau)\}_{(A_i)\in\mathcal{D}}$.

Proof. The proofs for the general and the complete cases are the same, so we treat only the latter. All is straightforward except that for any complete X and any family $\{\tilde{\psi}_{(A_i)} : (\ell^\infty,\tau) \to X\}_{(A_i)\in\mathcal{D}}$ of continuous linear maps such that

$\tilde{\psi}_{(A_i)} = \tilde{\psi}_{(B_i)} \circ \tilde{\Phi}_{(A_i)(B_i)}$ whenever $(A_i) \leq (B_i)$ in $\mathcal{D}$, there exists unique continuous linear $\tilde{\psi} : L(\Sigma) \to X$ such that $\tilde{\psi}_{(A_i)} = \tilde{\psi} \circ \tilde{\Phi}_{(A_i)}$ for all $(A_i) \in \mathcal{D}$. But $\tilde{\psi}$ can be defined by setting, for any $A \in \Sigma$, $\psi(A) = \psi_{(A_i)}(a)$ where $(A_i) \in \mathcal{D}$ is chosen so that $A = \cup A_n$ over $n \in a$, some a. That this definition does not depend on the choice of the partition (A_i) refining A and that ψ is a measure uniquely satisfying $\tilde{\psi}_{(A_i)} = \tilde{\psi} \circ \tilde{\Phi}_{(A_i)}$ is easy to establish, and the proof is complete.

It is easy to see that $(S(\Sigma),\tau)$ is the inductive limit of spaces (m_0,τ) as indexed above; however, $L(\Sigma)$ is not, in general, the inductive limit of spaces (ℓ^∞,τ) since the ranges $\tilde{\Phi}_{(A_i)}(\ell^\infty)$, $(A_i) \in \mathcal{D}$, do not cover $L(\Sigma)$. Thus, the distinction between direct limits and inductive limits is important in the complete case, and since direct limits are defined in terms of universal mapping properties, the next result is an immediate consequence of the first.

Corollary 2. For every X (respectively, complete X), the mapping $\tilde{\Phi} \longrightarrow (\tilde{\Phi}\circ\tilde{\Phi}_{(A_i)})_{(A_i)\in\mathcal{D}}$ is a linear isomorphism of $L((S(\Sigma),\tau),X)$ (respectively, $L(L(\Sigma),X))$ onto the subspace of the product over $\mathcal{D}$ of $L((m_0,\tau),X)$ (respectively, $L((\ell^\infty,\tau),X))$ consisting of all $(\tilde{\psi}_{(A_i)})_{(A_i)\in\mathcal{D}}$ such $\tilde{\psi}_{(A_i)} = \tilde{\psi}_{(B_i)} \circ \tilde{\Phi}_{(A_i)(B_i)}$ when $(A_i) \leq (B_i)$. That is, $L((S(\Sigma),\tau),X)$ is an inverse (or projective) limit of spaces $L((m_0,\tau),X)$, and $L(L(\Sigma),X)$ is an inverse limit of spaces $L((\ell^\infty,\tau),X)$. In each case, the inverse system of maps is the one naturally induced by the corresponding direct system in Theorem 1.

The isomorphisms of Corollary 2 are algebraic, but one of them is also topological if the topology s of simple convergence is used.

Theorem 3. For any X, the mapping $\tilde{\Phi} \longrightarrow (\tilde{\Phi} \circ \tilde{\Phi}_{(A_i)})_{(A_i)\in\mathcal{D}}$ (respectively, $\Phi \longrightarrow (\tilde{\Phi} \circ \Phi_{(A_i)})_{(A_i)\in\mathcal{D}})$ is a linear homeomorphism of $L_s((S(\Sigma),\tau),X)$ (respectively, $ca_s(\Sigma,X)$ with the topology s of convergence on each $A \in \Sigma$) onto the closed subspace of the product $\prod_{\mathcal{D}} L_s((m_0,\tau),X)$ (respectively, $\prod_{\mathcal{D}} ca_s(\mathcal{P}(\mathbb{N}),X))$ consisting of all $(\tilde{\psi}_{(A_i)})$ (respectively, $(\psi_{(A_i)})$) such that $\tilde{\psi}_{(A_i)} = \tilde{\psi}_{(B_i)} \circ \tilde{\Phi}_{(A_i)(B_i)}$ (respectively, $\psi_{(A_i)} = \tilde{\psi}_{(B_i)} \circ \Phi_{(A_i)(B_i)})$ when $(A_i) \leq (B_i)$.

Proof. The linear isomorphism $L((S(\Sigma),\tau),X) \simeq ca(\Sigma,X)$ is clearly a homeomorphism in the topologies of simple convergence, and it suffices to consider the mapping $\Phi \longrightarrow (\tilde{\Phi} \circ \Phi_{(A_i)})$. But because for any $A \in \Sigma$, there exists $(A_i) \in \mathcal{D}$ such that $A = \cup A_n$ over $n \in a$, some a, it is easy to see that $\lim_\alpha \Phi_\alpha(A) = \Phi(A)$ for every $A \in \Sigma$ iff $\lim_\alpha \tilde{\Phi}_\alpha(\Phi_{(A_i)}(a)) = \tilde{\Phi}(\Phi_{(A_i)}(a))$ for all $a \in \mathcal{P}(\mathbb{N})$ and $(A_i) \in \mathcal{D}$. It remains to show that if $\lim_\alpha \tilde{\Phi}_\alpha(\Phi_{(A_i)}(a)) = \psi_{(A_i)}(a)$ for some net$\{\Phi_\alpha\}$ from $ca(\Sigma,X)$, some $(\psi_{(A_i)}) \in \prod_{\mathcal{D}} ca(\mathcal{P}(\mathbb{N}),X)$, and every $a \in \mathcal{P}(\mathbb{N})$ and $(A_i) \in \mathcal{D}$, then $\psi_{(A_i)} = \tilde{\psi}_{(B_i)} \circ \Phi_{(A_i)(B_i)}$ when $(A_i) \leq (B_i)$. But, taking limits, this follows from the fact that $\Phi_{(A_i)} = \tilde{\Phi}_{(B_i)} \circ \Phi_{(A_i)(B_i)}$ when $(A_i) \leq (B_i)$.

Corollary 4. Let $K \subseteq ca(\Sigma,X)$ for any σ-algebra Σ on Ω and any locally convex X. Then K is relatively s-compact where s is the topology of simple convergence (on each $A \in \Sigma$) iff for every partition of Ω into a pairwise disjoint sequence (A_i) of members of Σ, the restriction of K to the sub-σ-algebra of Σ generated by the A_i's is relatively s-compact.

3. Compactness in $ca(\Sigma, X)$

Relative s-compactness can be characterized with the aid of Corollary 4.

Lemma 5. Let $K \subseteq ca(\mathcal{P}(\mathbb{N}), X)$, and assume that

a) K is uniformly countably additive, and

b) $K(a)$ is relatively compact for all $a \in \mathcal{P}(\mathbb{N})$.

Then K is relatively s-compact.

Proof. Consider a net $\{\Phi_\alpha\}$ from K. Let $\mathcal{F}(\mathbb{N})$ be the ring of all finite subsets of $\mathbb{N}$ and (a_n) an enumeration of it. An application of (b) gives subnets $\{\Phi_{n_\alpha}\}$ of $\{\Phi_\alpha\}$ such that $\{\Phi_{n_\alpha}(a_n)\}$ is convergent for each n and $\{\Phi_{n_\alpha}\}$ is a subnet of $\{\Phi_{k_\alpha}\}$ when $k < n$. The latter condition along with the additivity of each Φ_α insures that an additive map Φ is defined on $\mathcal{F}(\mathbb{N})$ by $\Phi(a_n) = \lim_{n_\alpha} \Phi_{n_\alpha}(a_n)$. Let $b_k \subseteq b_{k+1}$ for each k where (b_k) is a sequence from $\mathcal{F}(\mathbb{N})$. If $(\Phi(b_k))$ can be proved to be a convergenct sequence in X, then Φ may be assumed to have been uniquely extended to a countably additive map on $\mathcal{P}(\mathbb{N})$, also called Φ [8; 6.4]. To see that $(\Phi(b_k))$ is convergent, let $b = \cup b_k$ and apply (a) to choose, for each neighborhood U of 0 in X, an integer $n(U)$ such that $\psi(b) - \psi(b_k) \in U$ for each $\psi \in K$ and each $k \geq n(U)$. From the subnet of $\{\Phi_\alpha(b_{n(U)})\}$ which converges to $\Phi(b_{n(U)})$, pick a term $\Phi_{\alpha(U)}(b_{n(U)})$ such that $\Phi(b_{n(U)}) - \Phi_{\alpha(U)}(b_{n(U)}) \in U$. Apply (b) to find some $x \in X$ as the limit of a subnet of the net $\{\Phi_{\alpha(U)}(b)\}$ over the (directed) neighborhood system of 0 in X. To see that $x = \lim_k \Phi(b_k)$, fix V and select $U \subseteq V$ such that $x - \Phi_{\alpha(U)}(b) \in V$. Then for $k \geq n(U)$, $b_k = a_m$ and $b_{n(U)} = a_r$ for some m and r, and, since one of $\{\Phi_{m_\alpha}(a_m)\}$ and $\{\Phi_{r_\alpha}(a_r)\}$ is a subnet of the other, Φ_β may be chosen from one of these so that $\Phi(a_m) - \Phi_\beta(a_m) \in U$ and $\Phi(a_r) - \Phi_\beta(a_r) \in U$. But $\Phi_\beta(b_{n(U)}) - \Phi_\beta(b_k) \in 2U$ since $\Phi_\beta \in K$ and $k \geq n(U)$. Thus, $\Phi(b_{n(U)}) - \Phi(b_k) \in 4U$. So for $k \geq n(U)$,

$$x - \Phi(b_k) = [x - \Phi_{\alpha(U)}(b)] + [\Phi_{\alpha(U)}(b) - \Phi_{\alpha(U)}(b_{n(U)})]$$

$$+ [\Phi_{\alpha(U)}(b_{n(U)}) - \Phi(b_{n(U)})] + [\Phi(b_{n(U)}) - \Phi(b_k)]$$

$$\in V + U + U + 4U \subseteq 7V .$$

This proves that $x = \lim_k \Phi(b_k)$. So Φ may be assumed defined and countably additive on $P(\mathbb{N})$. To now see that $\{\Phi_\alpha\}$ s-clusters at Φ, it suffices to prove that $\{\Phi_\alpha(a)\}$ clusters at $\Phi(a)$ for an infinite subset a of $\mathbb{N}$. To this end, fix α_0, let V be as above, and apply (a) to find a_m from the original enumeration (a_n) of $F(\mathbb{N})$ such that $\psi(a) - \psi(a_m) \in V$ for all $\psi \in K$ and $\Phi(a) - \Phi(a_m) \in V$. Then choosing $m_\alpha > \alpha_0$ such that $\Phi(a_m) - \Phi_{m_\alpha}(a_m) \in V$ gives

$$\Phi(a) - \Phi_{m_\alpha}(a) = [\Phi(a) - \Phi(a_m)] + [\Phi(a_m) - \Phi_{m_\alpha}(a_m)] + [\Phi_{m_\alpha}(a_m) - \Phi_{m_\alpha}(a)] \in 3V ,$$

thereby establishing the relative s-compactness of K.

Consider the following conditions on $K \subseteq ca(\Sigma,X)$.

I. $K(\Sigma)$ is bounded.

II. K is uniformly countably additive.

III. $K(A)$ is relatively compact for each $A \in \Sigma$.

IV. $K(\Sigma')$ is relatively compact for every sub-σ-algebra Σ' of Σ generated by a sequence of pairwise disjoint members of Σ which partitions Ω.

Lemma 6. If K is relatively s-compact in $ca(\Sigma,X)$, then K satisfies (I), (II), and (III). If $\Sigma = P(\mathbb{N})$, then also $K(P(\mathbb{N}))$ is relatively compact.

Proof. (I) and (II) are satisfied by K iff $\tilde{K}$ is equicontinuous [8; 4.3]. But $\tilde{K}$ is equicontinuous iff $E\circ\tilde{K}$ is equicontinuous for every equicontinous $E \subseteq X^*$ iff [8; 11.6] $E\circ\tilde{K}$ is relatively weak $*$-compact for equicontinuous $E\subseteq X^*$. Let $\{x_\alpha^* \circ\tilde{\Phi}_\alpha\}$ be any net from $E\circ\tilde{K}$ for equicontinuous $E \subseteq X^*$. By the assumption on K and the Alaoglu-Bourbaki theorem applied to E, there exist $\tilde{\Phi} \in L((S(\Sigma),\tau),X)$ and $x^* \in X^*$ such that, without loss of generality, $\tilde{\Phi} = \lim_\alpha \tilde{\Phi}_\alpha$ and $x^* = \lim_\alpha x_\alpha^*$ in the s and weak $*$-topologies respectively. To see that $x^*\circ\tilde{\Phi} = \lim_\alpha x_\alpha^*\circ\tilde{\Phi}_\alpha$ in the weak $*$-topology, fix $\varepsilon > 0$ and $f \in S(\Sigma)$ and choose U, a neighborhood of 0 in X, such that $|E(U)| < \varepsilon$. Choose α_0 such that $\tilde{\Phi}(f) - \tilde{\Phi}_\alpha(f) \in U$ for $\alpha > \alpha_0$. Now choose $\alpha_1 > \alpha_0$ such that $|x^*(\tilde{\Phi}(f)) - x_\alpha^*(\tilde{\Phi}(f))| < \varepsilon$ for $\alpha > \alpha_1$. Then for $\alpha > \alpha_1$,

$$|x^*(\tilde{\Phi}(f) - x_\alpha^*(\tilde{\Phi}_\alpha(f))| \le |x^*(\tilde{\Phi}(f)) - x_\alpha^*(\tilde{\Phi}(f))| + |x_\alpha^*(\tilde{\Phi}(f)) - x_\alpha^*(\tilde{\Phi}_\alpha(f))|$$

$$< \varepsilon + \varepsilon = 2\varepsilon .$$

So (I) and (II) hold for K, and it is clear that (III) holds. Now assume that $\Sigma = P(\mathbb{N})$ and consider a net $\{\Phi_\alpha(a_\alpha)\}$ from $K(P(\mathbb{N}))$. Since the range, $\chi(P(\mathbb{N}))$, of the universal measure χ on $P(\mathbb{N})$ must be τ-compact [8; sec. 5], we may assume that $\chi_a = \lim_\alpha \chi_{a_\alpha}$ in topology τ for some $a \in P(\mathbb{N})$. We may also assume that $\Phi = \lim_{\alpha} \Phi_{\alpha}$ in the s-topology on $ca(\mathcal{P}(\mathbb{N}),X)$ since K is relatively s-compact. But

$$\Phi(a) - \Phi_{\alpha}(a_{\alpha}) = [\Phi(a) - \Phi_{\alpha}(a)] + [\tilde{\Phi}_{\alpha}(\chi_{a}) - \tilde{\Phi}_{\alpha}(\chi_{a_{\alpha}})]$$

and since $\tilde{\Phi}_{\alpha} \in \tilde{K}$, an equicontinuous set by the above, it follows that $\Phi(a) = \lim_{\alpha} \Phi_{\alpha}(a_{\alpha})$.

Theorem 7. Let Σ be any σ-algebra of subsets of Ω, let X be any locally convex space, and let $K \subseteq ca(\Sigma,X)$. Then (1), (2), (3), (4), and (5) are equivalent. If X is complete in its Mackey topology, then (6) and (7) may be added to the list of equivalent statements.

1. K is relatively s-compact
2. K satisfies conditions (II) and (III).
3. K satisfies conditions (I), (II), and (IV).
4. $\tilde{K}$ is relatively compact in $L_s((S(\Sigma),\tau),X)$.
5. $\tilde{K}$ is equicontinuous in $L((S(\Sigma),\tau),X)$ and K satisfies condition (III).
6. $\tilde{K}$ is relatively compact in $L_s(L(\Sigma),X)$.
7. $\tilde{K}$ is equicontinuous in $L(L(\Sigma),X)$ and K satisfies condition (III).
8. $\tilde{K}$ is relatively compact in $L_c(L(\Sigma),X)$. (c = topology of uniform convergence on compacta)

Proof. Statements (1) and (4) are equivalent via the linear homeomorphism $ca(\Sigma,X) \simeq L((S(\Sigma),\tau),X)$, and they imply (3) by Lemma 6 and Corollary 4. Clearly, (3) implies (2). Moreover, (2) implies (1) by Corollary 4 and Lemma 5. Finally, (3) implies (5) by [8; 4.3], which also insures that (5) implies (2). If X is complete in its Mackey topology, then (5) and (7) are equivalent, and (6) implies (4). To see that (6) follows from (4), assume (4), and let $\{\tilde{\Phi}_{\alpha}\}$ be any net in $\tilde{K} \subseteq L(L(\Sigma),X)$. We may assume that $\lim_{\alpha} \tilde{\Phi}_{\alpha} = \tilde{\Phi}$ in $L_s((S(\Sigma),\tau),X)$. Let $F \in L(\Sigma)$, and let U be any neighborhood of 0 in X. Since $S(\Sigma)$ is τ-dense in $L(\Sigma)$, there is $f \in S(\Sigma)$ such that $\tilde{\Phi}(F) - \tilde{\Phi}(f) \in U$ and $\tilde{K}(F - f) \subseteq U$, the latter by the equicontinuity of $\tilde{K}$ ((5) and (7) are equivalent). So if α_0 is chosen such that $\tilde{\Phi}(f) - \tilde{\Phi}_{\alpha}(f) \in U$ for $\alpha \geq \alpha_0$, then

$$\tilde{\Phi}(F) - \tilde{\Phi}_{\alpha}(F) = [\tilde{\Phi}(F) - \tilde{\Phi}(f)] + [\tilde{\Phi}(f) - \tilde{\Phi}_{\alpha}(f)] + [\tilde{\Phi}_{\alpha}(f) - \tilde{\Phi}_{\alpha}(F)] \in 3U$$

for $\alpha \geq \alpha_0$. Finally, (8) may be added to the list by virtue of III. 4.5 of [15].

Among the applications of Theorem 7 are a Nikodym boundedness theorem and a Vitali-Hahn-Saks theorem for measures on a σ-algebra Σ taking values in a locally convex space X.

Theorem 8 (Nikodym). Let $K \subseteq ca(\Sigma,X)$. If $K(A)$ is bounded for every $A \in \Sigma$, then $K(\Sigma)$ is bounded.

Proof. Assume K is s-bounded but not uniformly bounded. Then there are sequences (Φ_n) from K and (A_n) from Σ and a semi-norm p from a family of semi-norms defining the topology of X such that $\lim_n p(\Phi_n(A_n)) = \infty$. Let $\alpha_n = \sqrt{p(\Phi_n(a_n))}$ and $\psi_n = \alpha_n^{-1}\Phi_n$ for each n. Then $\lim_n \psi_n(A) = 0$ for every $A \in \Sigma$ since $\Phi_n \in K$ and K is s-bounded. But then $\{\psi_n\}$ is a relatively s-compact set and so, by Theorem 7, is uniformly bounded, a contradiction since $\lim_n p(\psi_n(A_n)) = \infty$.

Theorem 9. (Vitali-Hahn-Saks). If $\lim_n \Phi_n(A)$ exists for a sequence (Φ_n) from $ca(\Sigma,X)$ and every $A \in \Sigma$, then (Φ_n) is uniformly countably additive.

Proof (after [7; I.4.8]). It suffices to prove that $\lim_k \Phi_n(A_k) = 0$ uniformly in n for every pairwise disjoint sequence (A_k) from Σ. If not, then after taking subsequences and relabeling, we may assume that $\Phi_n(A_n) \notin 2U$ for some neighborhood U of 0 in X and some pairwise disjoint sequence (A_n) from Σ. Taking another subsequence and again relabeling, we can assume that $\Phi_n(A_n) \notin 2U$ while $\Phi_n(A_{n+1}) \in U$ (since Φ_n is countably additive). Now let $\psi_n = \Phi_{n+1} - \Phi_n$. Then $\lim_n \psi_n(A) = 0$ for $A \in \Sigma$ since $\lim_n \Phi_n(A)$ exists. So $\{\psi_n\}$ is relatively s-compact, hence is uniformly countably additive by Theorem 7. But this is a contradiction since $\psi_n(A_{n+1}) \notin U$, and so the proof is complete.

Remark. If X is a Banach space then uniform (strong) countable additivity of K and relative weak compactness of each $K(A)$, $A \in \Sigma$, are known [7; IV.2.4] to be necessary conditions for relative weak compactness of a subset K of $L^1(\mu,X)$ where μ is any non-negative member of $ca(\Sigma)$. These conditions are also sufficient [7; IV.2.1] for relative weak compactness under the assumption that X and X^* have the Radon-Nikodym property. That they are sufficient for relative s-compactness of K as a subset of $ca(\Sigma,X_w)$, where X_w is X with its weak topology, is immediate from Theorem 7. If a net $\{\Phi_\alpha\}$ from $K \subseteq L^1(\mu,X)$ converges in the s-topology in $ca(\Sigma,X_w)$ to Φ, it is easy to see that Φ is μ-continuous, that Φ is of bounded variation, and, because X has the Radon-Nikodym property, that $\Phi \in L^1(\mu,X)$. It can then be shown that $\{\Phi_\alpha\}$ weakly converges to Φ in $L^1(\mu,X)$ by a slight modification of the argument in the last part of the proof of IV.2.1 in [7].

4. The Mackey topology $\tau(B(\Sigma), ca(\Sigma))$

Since $S(\Sigma) \subseteq B(\Sigma) \subseteq L(\Sigma)$ and both of $L(\Sigma)$ and $(S(\Sigma), \tau)$ are Mackey spaces [8; 4.5 and 11.8], the restriction of τ from $L(\Sigma)$ to $B(\Sigma)$ is the Mackey topology $\tau(B(\Sigma), ca(\Sigma))$. The purpose of this section is to describe τ in $B(\Sigma)$ and to give a recipe for a τ-neighborhood of 0 in $B(\Sigma)$ which is mapped to a relatively weakly compact set by $\tilde{\Phi}$ for $\Phi \in ca(\Sigma, X)$ when X is a Banach space.

As in section 2, $\mathcal{D}$ is a cofinal family of partitions of Ω into pairwise disjoint sequences (A_i) of members of Σ. Let $w \in c_0^+$ mean that $w \in c_0$ ($\lim_i w(i) = 0$) and $w(i) > 0$ for all i. For $w \in c_0^+$ and $(A_i) \in \mathcal{D}$, let

$$U(w, (A_i)) = \{f \in B(\Sigma) : ||f\Sigma w(i)\chi_{A_i}|| \leq 1\}$$

where $\Sigma w(i)\chi_{A_i}$ is the countably-valued function with value $w(i)$ at every point in A_i and the norm is the sup norm.

<u>Theorem 10</u>. The family of closed convex sets $\overline{co}(\bigcup_{(A_i)\in\mathcal{D}} U(w_{(A_i)}, (A_i)))$ where for each $(A_i) \in \mathcal{D}$, $w_{(A_i)} \in c_0^+$, constitute a neighborhood base at 0 for the Mackey topology $\tau(B(\Sigma), ca(\Sigma))$.

Proof. The reader who is familiar with Buck's strict topology on ℓ^∞ and with arrow-theoretic diversions will find that this follows from Theorem 1. However, an alternate proof, more in the spirit of the rather delicate proof of upcoming Theorem 11, is given here.

The closed convex sets described in the theorem do determine a locally convex topology ρ on $B(\Sigma)$. To see that τ is finer than ρ, it suffices [8; 1.6] to show that χ is ρ-countably additive. For sequence (B_i) of pairwise disjoint members of Σ, let $B = \cup B_i$, let A_1 be the complement of B, and let $A_k = B_{k-1}$ for $k > 1$. If $(A_k) \in \mathcal{D}$, and if for any $w \in c_0^+$, k_0 is chosen so that $w(k) \leq 1$ for $k > k_0$, then $||(\chi_B - \sum_{j\leq k} \chi_{B_j})\Sigma w(i)\chi_{(A_i)}|| \leq 1$ for $k \geq k_0$, and so χ is ρ-countably additive. Otherwise, a member of $\mathcal{D}$ refining (A_k) can be similarly used.

To prove that ρ is finer than τ, it suffices to produce, for each τ-neighborhood U of 0 and each $(A_i) \in \mathcal{D}$, a $w \in c_0^+$ such that $U(w, (A_i)) \subseteq U$. It may be assumed [8; 11.6] that $U = K^0$, the polar in $B(\Sigma)$ of a uniformly countably additive and uniformly bounded subset K of non-negative members of $ca(\Sigma)$. Let $(A_i) \in \mathcal{D}$, and assume that $\mu(\Omega) < M$ for all $\mu \in K$. Choose a strictly increasing sequence (N_k) of positive integers such that $\sum_{i\geq N_k} \mu(A_i) < 4^{-2(k+1)}$ uniformly for $\mu \in K$. Define $w_{(A_i)} \in c_0^+$ by $w_{(A_i)}(n) = 3M$ if $n < N_1$, and $w_{(A_i)}(n) = 4^{-k}$ if $N_{k-1} \leq n < N_k$ for $k > 1$. If $f \in U(w_{(A_i)}, (A_i))$, then for all $\mu \in K$

$$\left|\int f d\mu\right| \leq \int |f| d\mu = \Sigma \int_{A_i} |f| d\mu$$

$$\leq \frac{1}{3M}\mu(\bigcup_{i<N_1} A_i) + \sum_{k\geq 2} 4^k \mu(\bigcup_{N_{k-1}\leq i<N_k} A_i)$$

$$< \frac{1}{3} + \sum_{k\geq 2} 4^{-k} < 1,$$

which establishes that ρ is finer than τ and completes the proof.

If X is a Banach space and $\Phi \in ca(\Sigma, X)$, then $||\Phi||$ is the semi-variation of Φ defined by

$$||\Phi||(A) = \sup\{|x^* \circ \Phi|(A) : x^* \in X^*, ||x^*|| \leq 1\}$$

where $|x^* \circ \Phi|$ is the variation of the scalar-valued measure $x^* \circ \Phi$. Note that $\{|x^* \circ \Phi| : x^* \in X^*, ||x^*|| \leq 1\}$ is uniformly countably additive by [8; 4.3] since it is clearly τ-equicontinuous in $ca(\Sigma) \simeq L(\Sigma)^*$.

<u>Theorem 11</u>. Let X be a Banach space and $\Phi \in ca(\Sigma, X)$. For each $(A_i) \in \mathcal{D}$, choose a strictly increasing sequence (N_k) of positive integers such that $||\Phi||(\bigcup_{i\geq N_k} A_i) < 2^{-(2k+3)}$ and define $w_{(A_i)} \in c_0^+$ by $w_{(A_i)}(n) = 1$ for $n < N_1$ and $w_{(A_i)}(n) = 2^{-k}$ for $N_{k-1} \leq n < N_k$ and $k \geq 2$. Let $U = \overline{co}(\bigcup_{(A_i)\in\mathcal{D}} U(w_{(A_i)}, (A_i)))$. Then U is a $\tau(B(\Sigma), ca(\Sigma))$-neighborhood of 0 in $B(\Sigma)$, and $\tilde{\Phi}(U) = \{\int f d\Phi : f \in U\}$ is relatively weakly compact in X.

Proof. The first conclusion follows from Theorem 10. It is the relative weak compactness of $\tilde{\Phi}(U)$ which must be established. According to [8; 2.2 and 2.4], $(S(\Sigma), \tau)$ has a fundamental sequence of bounded sets since the τ-bounded and the sup norm bounded subsets of $S(\Sigma)$ are the same, and τ is the finest locally convex topology on $S(\Sigma)$ which agrees with τ on τ-bounded subsets. So $(S(\Sigma), \tau)$ is what has been called in [13] a (gDF) (generalized DF) space. According to [12; 2.4], $(L(\Sigma),\tau)$ and $(B(\Sigma),\tau)$ are then (gDF) spaces. The τ-bounded subsets of $B(\Sigma)$ are those which are bounded in sup norm. As observed in [12; 3.1], (gDF) spaces are quasinormable [9, Def. 4, p. 106]. So $(B(\Sigma),\tau)$ is quasinormable, and, according to [8; 10.6], τ-bounded subsets of $B(\Sigma)$ are mapped by $\tilde{\Phi}$ into relatively weakly compact sets. Thus, an application of a theorem of Grothendieck [9, Cor. 1 of Th. 11, p. 114] yields a τ-neighborhood V of 0 in $B(\Sigma)$ such that $\tilde{\Phi}(V)$ is relatively weakly compact. Moreover, V may be taken to be $\bigcap_{m\in\mathbb{N}}(m\mathcal{B}+\frac{1}{m}\tilde{\Phi}^{-1}(\mathcal{B}_X))$ where $\mathcal{B}$ and $\mathcal{B}_X$ are the closed unit balls in $B(\Sigma)$ and X respectively. That this is so is a special case of results in the preprint [14], and arguments are given here for the convenience of the reader: the indicated intersection is a τ-neighborhood of 0 by [11; Lemma 1] since $\tilde{\Phi}$ is continuous and τ is the finest locally convex topology on $B(\Sigma)$ agreeing with τ on $\mathcal{B}$, and the relative weak compactness of $\tilde{\Phi}(V)$ follows from that of $\tilde{\Phi}(\mathcal{B})$ and Lemma 1 in [10; Ch. 5, Part 4].

Thus it remains to show that U as defined in the statement of the theorem is a subset of V as defined above. It suffices to prove for fixed m, $(A_i)\in\mathcal{D}$ and $w_{(A_i)}\in c_0^+$ as defined in the statement of the theorem that

$$U(w_{(A_i)},(A_i))\subseteq m\mathcal{B}+\frac{1}{m}\tilde{\Phi}^{-1}(\mathcal{B}_X).$$

Set $B_1=\bigcup_{i<N_1}A_i$ and $B_k=\bigcup_{N_{k-1}\leq i<N_k}A_i$ for $k>1$ (where (N_k) is as defined in the statement of the theorem). Choose k_0 minimal such that $2^{-k_0}<1/m$. Let $A=\bigcup_{i<k_0}B_i$ and $B=\bigcup_{i\geq k_0}B_i$. For $f\in U(w_{(A_i)},(A_i))$, let $g_m=f\chi_A$ and $h_m=f\chi_B$. Then $f=g_m+h_m$, and $g_m\in m\mathcal{B}$ since for $x\in A$, $|f(x)|\leq 2^{k_0-1}\leq m$. The proof will be complete provided $h_m\in\frac{1}{m}\tilde{\Phi}^{-1}(\mathcal{B}_X)$; that is, provided $\|\int h_m d\Phi\|<\frac{1}{m}$. But for $x^*\in X^*$ with $\|x^*\|\leq 1$,

$$\begin{aligned}|\langle\int h_m d\Phi, x^*\rangle| &= |\int h_m d(x^*\circ\Phi)| \\ &\leq \int|h_m|d|x^*\circ\Phi| \\ &= \sum_{k\geq k_0}\int_{B_k}|f|d|x^*\circ\Phi| \\ &\leq \sum_{k\geq k_0}2^k|x^*\circ\Phi|(B_k) \\ &\leq \sum_{k\geq k_0}2^k\|\Phi\|(B_k) \\ &\leq \sum_{k\geq k_0}2^k\cdot 2^{-(2(k-1)+3)}=2^{-k_0}<\frac{1}{m}.\end{aligned}$$

Corollary 12. Let X be a Banach space. Let $\tilde{\Phi} : B(\Sigma) \to X$ be any bounded linear operator and let $\Phi : \Sigma \to X$ be defined by $\Phi(A) = \tilde{\Phi}(\chi_A)$. The following are equivalent.

1. Φ is countably additive.
2. $\tilde{\Phi}(U)$ is relatively weakly compact for some $\tau(B(\Sigma), ca(\Sigma))$-neighborhood U of 0 (in which case, U may be chosen as in Theorem 11).

Proof: If condition (2) holds, then Φ is continuous for $\tau = \tau(B(\Sigma), ca(\Sigma))$, and so Φ is countably additive. The converse is Theorem 11.

Remarks. The corollary is the countably additve analogue of a theorem from [5] (or see [7; VI. 1.1]) which asserts that Φ is strongly bounded iff $\tilde{\Phi}(\mathcal{B})$ is relatively weakly compact for some bounded subset $\mathcal{B}$ of $B(\Sigma)$. If $\Sigma = \mathcal{P}(\mathbb{N})$ and X is the scalar field, then Theorem 11 gives the well known fact that there is no fastest converging absolutely convergent series. Because for each non-negative $\mu \in ca(\Sigma)$ there is a τ-projection P_μ on $L(\Sigma)$ such that $P_\mu(L(\Sigma)) \simeq (L^\infty(\mu), \tau(L^\infty(\mu), L^1(\mu)))$ (see [8; 11.5]), each of the results of this section has its $L^\infty(\mu)$-analogue.

References

1. R. G. Bartle, N. Dunford, and J. T. Schwartz, Weak compactness and vector measures, Canad. J. Math. 7 (1955), 289-305.

2. J. Batt, On weak compactness in spaces of vector-valued measures and Bochner integrable functions in connection with the Radon-Nikodym property of Banach spaces, Rev. Roumaine Math. Pures Appl. 19 (1974), 285-304.

3. J. K. Brooks and N. Dinculeanu, Conditional expectations and weak and strong compactness in spaces of Bochner integrable functions, J. Multivariate Analysis 9 (1979).

4. R. C. Buck, Bounded continuous functions on a locally convex space, Mich. J. Math. 5 (1958), 95-104.

5. J. Diestel, Applications of weak compactness and bases to vector measures and vectorial integration, Rev. Roumaine Math. Pures Appl. 18 (1973), 211-224.

6. ________, Remarks on weak compactness in $L_1(\mu,X)$, Glasgow Math. J. 18 (1977), 87-91

7. J. Diestel and J. J. Uhl, Jr., Vector Measures, Amer. Math. Soc. Math. Surveys 15, 1977, Providence, RI.

8. W. H. Graves, On the theory of vector measures, Amer. Math. Soc. Memoirs 195, 1977, Providence, RI.

9. A. Grothendieck, Sur les espaces (F) et (DF), Summa Brasil. Math. 3 (1954), 57-122.

10. ________, Topological Vector Spaces, Gordon and Breach, N.Y., London, Paris, 1973.

11. W. Roelcke, On the finest locally convex topology agreeing with a given topology on a sequence of absolutely convex sets, Math. Ann. 198 (1972), 57-80.

12. W. Ruess, On the locally convex structure of strict topologies, Math. Z. 153 (1977), 179-192.

13. ________, The strict topology and (DF) spaces, Functional Analysis: Surveys and Recent Results, ed. K.-D. Bierstedt and B. Fuchssteiner, Math. Studies 27, North Holland, New York, 1977.

14. ________, Generalized (DF) spaces and compactness of linear operators, to appear.

15. H. H. Schaefer, Topological Vector Spaces, Springer-Verlag, New York, Heidelberg, Berlin, 1971.

16. D. S. Sentilles, An L^1-space for Boolean algebras and semi-reflexivity of spaces $L^\infty(X,\Sigma,\mu)$, Trans. Amer. Math. Soc. 226 (1977), 1-37.

Contemporary Mathematics
Volume 2
1980

AXIOMATIC INFINITE SUMS - AN ALGEBRAIC APPROACH TO INTEGRATION THEORY

by

Denis Higgs

Abstract: The suggestion is made to base an algebraic approach to integration on the notion of a Σ-vector space, namely a vector space in which certain infinite sums are specified and obey the reasonable axioms. Σ-spaces, together with various related types of Σ-structure, are defined and a brief outline from the Σ-space point of view of integration in the simplest case of a real-valued function with respect to a non-negative real-valued measure is given.

§1. Introduction. An integral is customarily regarded as a linear function satisfying some further condition relating to topological, or occasionally order-theoretic, structure on the vector spaces involved. In a purely algebraic approach, such a further condition would be rather the preservation of additional algebraic structure, which we take to be Σ-structure: the specification of certain infinite sums obeying the reasonable postulates. Since we wish to work with vector spaces, it is unfeasible to require that all even countably infinite sums exist - our enriched vector spaces, although algebraic in nature, will only be partial

Copyright © 1980, American Mathematical Society

algebras as far as infinitary operations are concerned.

(An alternative approach, which avoids the use of partial operations but which is also comprised within the general scheme of ideas outlined below, is to model one's algebraic structures on $[0,\infty]$ rather than on $\mathbb{R}$ or $\mathbb{C}$. The resulting objects, called magnitude modules [2], [3], do admit all countable sums but possess some awkward features, for example subtraction is no longer available and the development inextricably involves order - a magnitude module tends to look something like the positive cone of an ordered vector space plus 'infinite' elements.)

Apart from its possible utility in integration and elsewhere in analysis, (where in particular it might conceivably provide a tool for the application of the methods of a suitable first-order logic), the notion of Σ-structure appears to be of interest in its own right. More detailed and extensive discussions of the notion will be presented in subsequent papers.

§2. Σ-structures. A Σ-monoid is a non-empty set A together with a partially defined operation Σ which assigns to certain 'summable' families a_i, $i \in I$, of elements of A a 'sum' $\sum_{i\in I} a_i$ in A in such a way that the following axioms are satisfied:

(I) If $I = \{k\}$ then $\sum_{i\in I} a_i$ exists and equals a_k;

(II) If $\sum_{i\in I} a_i$ exists and $f\colon I \to J$ is any function then $\sum_{f(i)=j} a_i$ exists for each $j \in J$, and $\sum_{j\in J}(\sum_{f(i)=j} a_i)$ exists and equals $\sum_{i\in I} a_i$;

(III) The same as (II) but with J finite and with the conditions "$\sum_{i \in I} a_i$ exists" and "$\sum_{f(i)=j} a_i$ exists for each $j \in J$" interchanged.

Conditions (II) and (III) here are (very nearly) the same as Théorème 2 and Proposition 3 respectively of Bourbaki [1], Chapitre 3, §5; thus every complete commutative Hausdorff topological monoid is a Σ-monoid under the definition of Σ given there. On the other hand, every Σ-monoid becomes a commutative monoid if we define $a_1 + a_2 = \sum_{i=1,2} a_i$, $0 = \sum_{i \in \phi} a_i$ (each infinite sum with all but finitely many terms equal to 0 then equals the usual finite sum of the non-zero terms).

A <u>Σ-group</u> is a Σ-monoid which, under the addition just defined, is a group (necessarily commutative) and in which, if Σa_i exists, so does $\Sigma(-a_i)$ (with $-(\Sigma a_i) = \Sigma(-a_i)$ necessarily). Every complete commutative Hausdorff topological group is a Σ-group (by the above, together with Proposition 6(4) in the same §5 of Bourbaki).

A <u>morphism</u> ('Σ-morphism') of Σ-monoids or Σ-groups A, B is a function $f: A \to B$ such that if Σa_i exists then $\Sigma f(a_i)$ exists and equals $f(\Sigma a_i)$. Σ-monoids (and likewise Σ-groups) form a category which is closed provided we define Σ on hom-sets by stipulating that $\sum_{i \in I} f_i$ exists iff, whenever $\sum_{j \in J} a_i$ exists, then also $\sum_{i,j} f_i(a_j)$ exists, with the value of Σf_i being given pointwise: $(\Sigma f_i)(a) = \Sigma f_i(a)$. Kuroš [4] considers categories which, roughly, are defined over the category of Σ-monoids (his condition I is stronger than our (II)).

Let A be a submonoid of a Σ-monoid B; then the Σ-structure on A <u>induced</u> by that on B is defined as follows: $\sum_{i\in I} a_i = a$ in A iff $\sum_{i\in I} a_i = a$ in B and $\sum_{i\in J} a_i \in A$ for all subsets J of I (this is the 'strongest' Σ on A for which A becomes a Σ-monoid such that the inclusion $A \to B$ is a Σ-morphism). Similarly for Σ-groups and the other types of Σ-structure defined below.

A <u>bimorphism</u> ('Σ-bimorphism') of Σ-monoids or Σ-groups A,B,C is a function $f\colon A\times B \to C$ such that if Σa_i and Σb_j exist then $\sum_{i,j} f(a_i,b_j)$ exists and equals $f(\Sigma a_i, \Sigma b_j)$.

A <u>Σ-ring</u> is a Σ-group which is also a monoid $(R,\cdot)$ with $R\times R \xrightarrow{\cdot} R$ a Σ-bimorphism. A (left) <u>Σ-module</u> over a Σ-ring R is a Σ-group A together with an action $R\times A \xrightarrow{\cdot} A$ (so that $(rs)a = r(sa)$, $1a = a$) which is a Σ-bimorphism. If we consider Σ-monoids here instead of Σ-groups we obtain the notions of <u>Σ-semiring</u> and <u>Σ-semimodule</u>; magnitude modules are Σ-semimodules (of a certain type) over the Σ-semiring $[0,\infty]$. Morphisms of Σ-rings etc. are just Σ-morphisms preserving the extra structure (multiplication, scalar multiplication).

The basic Σ-rings of analysis are $\mathbb{R}$ and $\mathbb{C}$ with Σ-structure (the 'usual' Σ-structure) given by the usual topology as in [1]; equivalently, $\Sigma a_i = a$ iff Σa_i is absolutely convergent with sum a. Σ-modules over $\mathbb{R}$ or $\mathbb{C}$ will be called <u>Σ-vector spaces</u>, or just <u>Σ-spaces</u>.

In connection with the results of [2], we remark that real Σ-spaces can be characterized as the Σ-groups admitting a unary operation h such that

(i) $h(a+b) = h(a) + h(b)$, and

(ii) if $\sum_{i \in I} a_i$ exists then $\sum h^{n+k_n}(a_i)$ exists and equals $\sum_{i \in I} a_i$, where $k_n, n \in N = \{1,2,3,...\}$, is any sequence of non-negative integers and the summation in $\sum h^{n+k_n}(a_i)$ is over $i \in I$, $m \in \{1,2,3,...,2^{k_n}\}$, and $n \in N$.

Notably, $\mathbb{R}$ can be defined as the free such Σ-group on one generator (the k_n's are required for the full Σ-structure on $\mathbb{R}$ to be obtained). Complex Σ-spaces, and $\mathbb{C}$ itself, have a similar description involving the obvious unary operation i as well as h.

§3. Σ-spaces and integration. Let S be a δ-ring of subsets of a set X and let M be the set of all real-valued functions on X which are of the form (M): $f = \sum_{n \in N} \alpha_n \chi_{E_n}$ where $\alpha_n \in \mathbb{R}$, $E_n \in S$, and the series is pointwise absolutely convergent. Then M is a real Σ-space with Σ given by pointwise absolute convergence.

The theorem of [3], as pointed out in the final paragraph there, has an implication for M(*) but it does not imply that M is freely generated as a Σ-space by the χ_E's, $E \in S$, subject to the relations $\chi_E = \Sigma\chi_{E_n}$, E the countable disjoint union of the sets $E_n \in S$. The reason for this is that the existence of the sums $\Sigma\chi_{E_n}$ by no means implies the existence of all the sums which in fact do exist in M. As suggested in [3], what we can say is that if

(*) In [3], S was a σ-ring but the proof goes through without change for S a δ-ring.

$\Sigma\alpha_n\chi_{E_n}$ and $\Sigma\beta_n\chi_{E_n}$ are two series of the form (M) which represent the same function then their equality is provable from the Σ-space axioms and the above-mentioned relations, together with the relevant summability statement (namely that asserting the summability of the intermediate series obtained by the process of [3] as a common refinement of the two series which arise when negative terms in the equation $\Sigma\alpha_n\chi_{E_n} = \Sigma\beta_n\chi_{E_n}$ are taken across to the other side).

The significance of this for the existence of integrals appears to be as follows. Let $\mu: S \to \mathbb{R}_+$ be a countably additive measure, let L consist of the functions f in M which can be expressed in the form (L): $f = \sum_{n\in N} \alpha_n\chi_{E_n}$ with $\Sigma\alpha_n\mu(E_n)$ absolutely convergent, and put $\overline{\mu}(f) = \Sigma\alpha_n\mu(E_n)$. That $\overline{\mu}: L \to \mathbb{R}$ is thereby well-defined may be seen in this way: - Let $\Sigma\alpha_n\chi_{E_n}$ and $\Sigma\beta_n\chi_{E_n}$ be two series of the form (L) which define the same function and in the proof (as above) that $\Sigma\alpha_n\chi_{E_n} = \Sigma\beta_n\chi_{E_n}$, replace each χ_E occuring by $\mu(E)$. Then the relations $\chi_E = \Sigma\chi_{E_n}$ used in the proof become the equations $\mu(E) = \Sigma\mu(E_n)$ which are equally true by virtue of the countable additivity of μ; also, on account of the summability of $\Sigma\alpha_n\mu(E_n)$ and $\Sigma\beta_n\mu(E_n)$, the intermediate series referred to in the preceding paragraph is easily seen to be transformed into a summable series. In consequence the whole proof is transformed into a valid proof that $\Sigma\alpha_n\mu(E_n) = \Sigma\beta_n\mu(E_n)$, as required. It is a very pleasing fact that if L is given the Σ-structure induced by that on M (see §2) then $\overline{\mu}$ is a Σ-space morphism (I am indebted to Ken Davidson for a proof of this).

A situation analogous to that just sketched arises if we look at Lang's regulated function approach to integration [5], Chapter X. Let S be a ring of subsets of a set X and let M_r be the set of all real-valued functions on X which are uniform limits of step functions based on S; equivalently M_r is the set of all f of the form (M_r): $f = \sum_{n \in N} \alpha_n \chi_{E_n}$ where $\alpha_n \in \mathbb{R}$, $E_n \in S$, and $\Sigma\alpha_n$ is absolutely convergent. Then M_r is a Σ-space with Σ given by uniform absolute convergence. By a decomposition technique somewhat similar to that used in [3], it can be shown that if $\Sigma\alpha_n\chi_{E_n}$ and $\Sigma\beta_n\chi_{E_n}$ are two series of the form (M_r) which determine the same function then their equality is provable from the Σ-space axioms and the relations $\chi_E = \sum_{n=1}^{k} \chi_{E_n}$ where E is the <u>finite</u> disjoint union of the sets $E_1,\ldots,E_k \in S$, together with the appropriate summability statement again. It now follows that any <u>finitely</u> additive measure $\mu: S \to \mathbb{R}_+$ extends to an integral (=Σ-space morphism) $\overline{\mu}: L_r \to \mathbb{R}$ where L_r bears the same relationship to M_r as L does to M in the previous situation (if it happens that $\mu(E)$ is bounded for $E \in S$ then the absolute convergence of the coefficient sums $\Sigma\alpha_n$ implies that $L_r = M_r$, as Σ-spaces).

Although these ideas on integration are in a very rudimentary form (for example, signed measures are not dealt with, still less vector-valued measures or integrands), I hope they suggest that Σ-structure on a vector space is a plausible alternative to topological or order-theoretic structure, an alternative for which, perhaps, the underlying logic is more explicit and displays a greater degree of unity.

References

1. Bourbaki, N., Éléments de mathématique, Livre III, Topologie Générale, Hermann, Paris, 1960.

2. Higgs, D., A universal characterization of $[0,\infty]$, Nederl. Akad. Wetensch, Proc. Ser. A. 81 (= Indag. Math. 40)(1978), 448-455.

3. Higgs, D., Characteristic functions freely generate measurable functions, to appear in Fundamenta Mathematicae.

4. Kuroš, A.G., Direct decomposition in algebraic categories, Amer. Math. Soc. Transl. Ser. 2, 27(1963), 231-255.

5. Lang, S., Analysis I, Addison-Welsey, Don Mills, 1968.

Contemporary Mathematics
Volume 2
1980

ON SOME CLASSES OF BANACH SPACES AND GENERALIZED HARMONIC ANALYSIS

by

Ka-Sing Lau*

§1. Introduction.

In classical harmonic analysis, two types of functions f are considered: (i) f can be represented as a trigonometric series, i.e.

$$f(t) = \sum_{k=-\infty}^{\infty} c_k e^{i\omega_k t}, \qquad t \in R,$$

(ii) $f \in L^2(R)$. Then by the Inversion theorem,

$$f(t) = \frac{1}{\sqrt{2\pi}} \int_{-\infty}^{\infty} c(\omega) e^{i\omega t}\, d\omega.$$

In terms of optics, the spectrum of the light signal f in (i) is made up of finite or countable number of sharp lines of intensity $|c_k|^2$ at the frequency ω_k. The light signal f in (ii) has continuous spectrum on the frequency band R. Note that in this case, for each fixed $h > 0$,

$$\lim_{t \to \pm\infty} \int_t^{t+h} |f(\tau)|^2 d\tau = 0.$$

*This research was partially supported by NSF-MCS-7903638.
Copyright © 1980, American Mathematical Society

This means that the energy emitted by the signal during a time interval of fixed length h approaches 0 as the interval advances to infinity on the time axis.

In the early twenty century, some physicists such as Rayleigh, Schuster, Taylor, were interested on the type of white light signals f (e.g. sunlight) that have continuous spectrum and infinite energy (i.e. $\int_{-\infty}^{\infty} |f(t)|^2\, dt = \infty$). The two classical approaches do not seem to explain the behavior of such light signals satisfactorily. Wiener felt that the difficulty stemmed from the limitation of the classical theory. Around the twenties, he began developing a "generalized" harmonic analysis that could cover signals f on R which are on one hand so irregular that their spectrum are not made up of sharp lines alone and on the other so lastingly vigorous that $\int_t^{t+h} |f(\tau)|^2\, d\tau \not\to 0$ as $t \to \infty$. The class of functions $\mathcal{W}^2(R)$ he considers is the set of Borel measurable functions f on R such that

$$\lim_{T\to\infty} \frac{1}{2T} \int_{-T}^{T} |f(t)|^2\, dt$$

exists and is finite [19]. In order to study the spectrum of the functions $f \in \mathcal{W}^2(R)$ and the covariance function

$$\phi(\tau) = \lim_{T\to\infty} \frac{1}{2T} \int_{-T}^{T} f(t+\tau)\overline{f(t)}dt,$$

Wiener introduced the following integrated Fourier transformation W(f) of f defined by

$$g(u) = \frac{1}{2\pi} \left(\int_{-\infty}^{-1} + \int_{1}^{\infty} \frac{f(t)e^{-itu}}{-it}\, dt + \int_{-1}^{1} f(t)\, \frac{e^{-itu}-1}{-it}\, dt\right). \tag{1.1}$$

(The last term on the right hand side of (1.1) guarantees integrability about the origin). Analogous to the Plancherel theorem in the L^2 case, he showed that for $f \in \mathcal{W}^2(R)$

$$\lim_{T\to\infty} \frac{1}{2T} \int_{-T}^{T} |f(t)|^2 \, dt = \lim_{h\to 0^+} \frac{1}{2h} \int_{-\infty}^{\infty} |g(u+h) - g(u-h)|^2 \, du. \qquad (1.2)$$

It has been found that the theory of generalized harmonic analysis is applicable to diverse areas of pure and applied mathematics. In particular, it was used to consider the problem of anti-aircraft fire control with radar during World War II and brought into the theory of prediction and filtering. For a detail account of this and its relationship with Kolmogorov's stochastic process, the reader may refer to [1], [4], [10], [12], [15].

The class of functions $\mathcal{W}^2(R)$ defined above is, however, not closed under addition, hence many functional analytic techniques are not applicable in the theory. To remedy this, Masani developed a nonlinear Banach graph theory to study $\mathcal{W}^2(R)$ and its closed subspaces [14], [15]. Yet another approach is to embed $\mathcal{W}^2(R)$ into a larger Banach space; a suitable one will be the Marcinkiewicz space $\mathcal{M}^2(R)$ [11]. This space had been considered by Bohr and Følner [3] and Bertrandias [2]. It is the purpose of the paper to report some recent results of the joint work of Lee and the author on this direction. The detail will appear elsewhere ([6], [7], [8]).

§2. The Marcinkiewicz Spaces.

For $1 \leq p < \infty$, let $\mathcal{M}^p(R)$ denote the class of complex valued Borel measurable functions f on R such that

$$||f||_{\mathcal{M}^p} = \overline{\lim_{T\to\infty}} \left(\frac{1}{2T}\int_{-T}^{T} |f|^p\right)^{\frac{1}{p}} < \infty$$

and let $\mathcal{W}^p(R)$ be the set of $f \in \mathcal{M}^p(R)$ such that

$$\lim_{T\to\infty} \left(\frac{1}{2T}\int_{-T}^{T} |f|^p\right)^{\frac{1}{p}}$$

exist. Roughly, the above norm estimates the average behavior of f for large T. It is easy to see that for any function $f \in L^p(R)$, $||f||_{\mathcal{M}^p} = 0$. By identifying functions whose difference has zero norm, $(\mathcal{M}^p(R), ||\cdot||)$ is a Banach space ([3], [9]).

Let B^pAP be the class of (Besicovitch) almost periodic functions, i.e. the $\mathcal{M}^p$-closure of the set of trigonometric polynomials $\sum_{k=1}^{n} a_k e^{it_k(\cdot)}$, $t_k \in R$. It is known that for $1 < p < \infty$, B^pAP is a nonseparable, reflexing Banach space $((B^pAP)^* = B^{p'}AP$, $\frac{1}{p} + \frac{1}{p'} = 1)$ and that $B^pAP \subseteq \mathcal{W}^p(R)$, hence we have

<u>Proposition 2.1</u>. Let $1 < p < \infty$, then $\mathcal{M}^p(R)$ contains a nonseparable reflexive subspace.

We can also show that

<u>Proposition 2.2</u>. Let $1 \leq p < \infty$, then $\mathcal{M}^p(R)$ contains a subspace isomorphic to ℓ^∞.

For the extremal structure of the unit sphere $S(\mathcal{M}^p(R))$ of $\mathcal{M}^p(R)$, we have

<u>Theorem 2.3</u>. Let $1 < p < \infty$. Then each norm 1 function f in $\mathcal{W}^p(R)$ is an extreme point of $S(\mathcal{M}^p(R))$. $S(\mathcal{M}^1(R))$ does not contain any extreme point.

In order to study the duality properties of $\mathcal{M}^p(R)$, we introduce the following auxilliary spaces;

$$M^p(R) = \{f : f \text{ is Borel measurable on } R, \|f\|_{M^p} = \sup_{T \geq 1} \left(\frac{1}{2T}\int_{-T}^{T} |f|^p\right)^{\frac{1}{p}} < \infty\}$$

and

$$I^p(R) = \{f \in M^p(R) : \lim_{T\to\infty} \left(\frac{1}{2T}\int_{-T}^{T} |f|^p\right)^{\frac{1}{p}} = 0\}.$$

It is easy to show that $M^p(R)$ is a Banach space and $I^p(R)$ is a closed subspace of $M^p(R)$.

<u>Theorem 2.4</u>. For $1 < p < \infty$

(i) $M^p(R)$ is the second dual of $I^p(R)$ and

(ii) $\mathcal{M}^p(R)$ is isometrically isomorphic to $M^p(R)/I^p(R)$.

It follows from Theorem 2.4 that for $1 < p < \infty$,

$$M^p(R)^* = I^p(R)^* \oplus I^p(R)^{\perp}$$

and $\mathcal{M}^p(R)^*$ is isometrically isomorphic to $I^p(R)^{\perp}$. The concrete representations of functionals on $I^p(R)$ and $\mathcal{M}^p(R)$ are given by

<u>Theorem 2.5</u>. Suppose that $1 < p < \infty$ and $\frac{1}{p} + \frac{1}{p'} = 1$.

(i) If $\ell \in I^p(R)^*$. Then there exists a $\psi \in M^{p'}(R)$ and a countably additive, positive, bounded regular Borel measure μ on $[1,\infty)$ such that for all $f \in I^p(R)$,

$$\langle \ell, f \rangle = \int_1^\infty \left(\frac{1}{2T}\int_{-T}^{T} f(t)\psi(t)dt\right) d\mu(T). \tag{2.1}$$

(ii) There exists a (norm) dense subset $D \subseteq \mathcal{M}^p(R)^*$ such that ℓ in D can be represented as in (2.1) with $\psi \in \mathcal{M}^{p'}(R)$ where μ

is a finite additive, positive, bounded regular Borel measure on $[1,\infty)$, which vanishes on bounded intervals.

§3. The Integrated Lipschitz Class.

For $1 \le p < \infty$, let $V^p(R)$ denote the class of complex valued Borel measurable functions on R such that

$$||g||_{V^p} = \overline{\lim_{\varepsilon \to 0}} \left(\frac{1}{2\varepsilon} \int_{-\infty}^{\infty} |g(u+\varepsilon) - g(u-\varepsilon)|^p du\right)^{\frac{1}{p}} < \infty.$$

This class of functions had been studied by Hardy and Littlewood in their investigation of fractional derivatives [5].

Let $g \in V^p(R)$ and let

$$\tilde{g} = \int_0^{\infty} e^{-t}(g - \tau_t g)dt$$

where τ_t is the translation operator defined by

$$(\tau_t f)(x) = f(x+t), \quad f \in V^p(R).$$

By using the theory of helixes in [13], we can show

Proposition 3.1. Let $1 < p < \infty$. Then for each $g \in V^p(R)$, there exists a $\tilde{g} \in L^p(R) \cap V^p(R)$ such that $||g - \tilde{g}||_{V^p} = 0$.

This amounts to saying that each equivalence class

$$[g] = \{f \in V^p(R): \ ||g - f||_{V^p} = 0\}$$

has a representation in $L^p(R)$.

Suppose g and its derivative g' are in $L^p(R)$, $1 < p < \infty$. Then

$$||g||_{V^p} = \overline{\lim_{h \to 0^+}} \left(\frac{1}{2h} \int_{-\infty}^{\infty} |g(u+h) - g(u-h)|^p \, du\right)^{\frac{1}{p}}$$

$$= \overline{\lim_{h \to 0^+}} (2h)^{\frac{p-1}{p}} \left|\left|\frac{1}{2h} (g(u+h) - g(u-h))\right|\right|_p$$

$$= 0 \cdot ||g'||_p = 0 .$$

Therefore, we have

Proposition 3.2. Let $1 < p < \infty$. If g and g' are in $L^p(R)$, then $||g||_{V^p} = 0$.

By identifying functions whose difference has zero norm, we can show that

Theorem 3.3. For $1 < p < \infty$, the normed linear space $V^p(R)$ is complete.

For the case $p = 1$, Nelson [16] showed that $V^1(R)$ is isometrically isomorphic to the space of bounded regular Borel measures on R, hence the properties of $V^1(R)$ are well known.

§4. The Integrated Fourier Transformation.

In proving the identity (1.2), Wiener introduced a fairly general form of Tauberian theorem which applies to functions in $W^2(R)$. In order to consider the Fourier transformation between $M^2(R)$ and $V^2(R)$, we prove another type of Tauberian theorem which applies to the limit supremum of functions at ∞.

Let M^+ denote the class of positive, Borel measurable functions on $[0,\infty)$ such that $\overline{\lim_{T\to\infty}} \frac{1}{T}\int_0^T f < \infty$.

Lemma 3.1. Let k be a positive continuous function on $[0,\infty)$. Assume that $\tilde{k}(t) = \sup_{x\geq t} k(x)$ is integrable and $C_1 = \int_0^\infty \tilde{k}(t)dt$. Then for all $f \in M^+$,

$$\overline{\lim_{T\to\infty}} \int_0^\infty f(Tt)k(t)dt \leq C_1 \overline{\lim_{T\to\infty}} \frac{1}{T}\int_0^T f(t)dt.$$

Moreover, C_1 is the best possible constant for $f \in M^+$.

<u>Lemma 3.2.</u> Let k be a positive continuous function on $[0,\infty)$, such that $\tilde{k}(t) = \sup_{x\geq t} k(x)$ is integrable. Suppose there is a t_0 satisfying $t_0k(t_0) = \max_{t>0} t\cdot k(t) (= C_2)$ and $k(t) \geq k(t_0)$ for all t in $[0,t_0]$. Then for all $f \in M^+$,

$$C_2 \overline{\lim_{T\to\infty}} \frac{1}{T}\int_0^T f(t)dt \leq \overline{\lim_{T\to\infty}} \int_0^\infty f(Tt)k(t)dt.$$

Moreover, C_2 is the best possible constant for $f \in M^+$.

It is easy to show that $M^2(R) \subseteq L^2(R, \frac{dt}{1+t^2})$, hence for $f \in M^2(R)$, the integral

$$\int_{-\infty}^{-1} + \int_1^\infty \frac{|f(t)|^2}{t^2} dt$$

exists. This implies that

$$\int_{-\infty}^{-1} + \int_1^\infty \frac{f(t)e^{-itu}}{-it} dt$$

converges in $L^2(R, \frac{dt}{1+t^2})$. Therefore, if $f \in M^2(R)$, we can define the integrated Fourier transformation $W(f) = g$ as

$$g(u) = \frac{1}{2\pi} \left(\int_{-\infty}^{-1} + \int_1^\infty f(t)\frac{e^{-itu}}{-it} dt + \int_{-1}^1 f(t)\frac{e^{-itu}-1}{-it} dt\right).$$

Now for $h > 0$,

$$(\tau_h g - \tau_{-h} g)(u) = \frac{1}{2\pi}\int_{-\infty}^\infty f(t)\frac{e^{ith}-e^{-ith}}{it} e^{-itu}dt$$

$$= \frac{1}{2\pi}\int_{-\infty}^\infty f(t)\frac{2\sin(ht)}{it} e^{-itu}dt.$$

Thus $(\tau_h g - \tau_{-h} g)$ is the Fourier transformation of

$$\sqrt{\frac{2}{\pi}}\, f(t)\frac{\sin(ht)}{t}$$

and the Plancherel theorem implies that

$$\frac{1}{2h}\int_{-\infty}^{\infty} |g(u+h) - g(u-h)|^2\, du = \frac{1}{h}\int_{-\infty}^{\infty} |f(t)|^2\, \frac{\sin^2 ht}{\pi t^2}\, dt.$$

Hence,

$$||W(f)||_{\mathcal{V}^2} = \overline{\lim_{h\to 0^+}}\, \frac{1}{h}\int_{-\infty}^{\infty} |f(t)|^2\, \frac{\sin^2 ht}{\pi t^2}\, dt$$

$$= \overline{\lim_{T\to\infty}} \int_{-\infty}^{\infty} |f(Tt)|^2\, \frac{\sin^2 t}{\pi t^2}\, dt.$$

Letting $k(t) = \frac{2\sin^2 t}{\pi t}$, $t > 0$ and let $\tilde{f}(t) = \frac{1}{2}\,(|f(t)|^2 + |f(-t)|^2)$, $t > 0$, Lemma 3.1 implies that $W(f) \in \mathcal{V}^2(R)$. Moreover, the lemma implies that $W(f) = 0$ for $f \in I^2(R)$. Since $\mathcal{M}^2(R) = M^2(R)/I^2(R)$, W induces a map from $\mathcal{M}^2(R)$ into $\mathcal{V}^2(R)$. Restating Wiener's theorem in [19], we have

<u>Theorem 4.3</u>. Let $f \in \mathcal{W}^2(R)$. Then $||W(f)||_{\mathcal{V}^2} = ||f||_{\mathcal{M}^2}$.

Our extension of Wiener's theorem is:

<u>Theorem 4.4</u>. The integrated Fourier transformation $W\colon\ \mathcal{M}^2(R) \to \mathcal{V}^2(R)$ is an isomorphism with

$$||W|| = \left(\int_0^{\infty} \tilde{k}(t)dt\right)^{\frac{1}{2}} \quad\text{and}\quad ||W^{-1}|| = \left(\max_{t\geq 0}\, t\cdot k(t)\right)^{-\frac{1}{2}}$$

where $k(t) = \frac{2\sin^2 t}{\pi t^2}$, $t > 0$ and $\tilde{k}(t) = \max_{x \geq t} k(x)$.

The two isomorphic constants are direct consequence of Lemma 4.1 and 4.2. Numerically, we find that $||W|| \approx 1.05$ and $||W^{-1}|| \approx 1.49$. The surjectivity of W is obtained as follows: for each $g \in \mathcal{V}^2(R)$, we may assume that $g \in L^2(R)$ (Proposition 3.1). Let $\check{g}$ denote the inverse Fourier transformation of g and let

$$f(t) = -i\sqrt{2\pi}\, t \cdot \check{g}(t). \tag{4.1}$$

By direct computation, we can show that $f \in \mathcal{M}^2(R)$ and $W(f) = g$.

For $p \neq 2$, we have the following:

Theorem 4.5. For $1 < p < 2$, the integrated Fourier transformation W defines a bounded linear operator form $\mathcal{M}^p(R)$ into $\mathcal{V}^{p'}(R)$ with

$$||W|| \leq \left(\int_0^\infty \tilde{k}(t)dt\right)^{\frac{1}{p}}$$

where $k(t) = \left|\frac{2\sin^p t}{\pi t^p}\right|$, $t > 0$.

§5. The Convolution Operators.

We call a function f in $M^p(R)$ regular if

$$\lim_{T \to \pm\infty} \int_T^{T+a} |f|^p = 0 \quad \text{for any } a > 0.$$

Let $M_r^p(R)$ denote the set of regular functions in $M^p(R)$ and let $\mathcal{M}_r^p(R) = M_r^p(R)/I^p(R)$. Then $\mathcal{M}_r^p(R)$ is a closed subspace of $\mathcal{M}^p(R)$ and $\mathcal{W}^p(R) \subseteq \mathcal{M}_r^p(R)$.

Let M denote the set of bounded regular Borel measure on R and let M_1 be the subspace of measures with bounded support. For $\mu \in M_1$, we define the convolution operator $\Phi_\mu : \mathcal{M}^p(R) \to \mathcal{M}^p(R)$ by

$$\Phi_\mu(f) = \mu * f, \qquad f \in \mathcal{M}^p(R).$$

Note that Φ_μ also defines an operator from $L^p(R)$ into $L^p(R)$. Restricting Φ_μ to $\mathcal{M}^p_r(R)$ yields

<u>Proposition 5.1</u>. For $\mu \in M_1$, the operator $\Phi_\mu : \mathcal{M}^p_r(R) \to \mathcal{M}^p_r(R)$ satisfies

$$\lim_{T\to\infty} \frac{1}{2T} \int_R |(\chi_{[-T,T]}\Phi_\mu - \Phi_\mu\chi_{[-T,T]})f|^p = 0.$$

By using this, we can show that for $f \in \mathcal{M}^p_r(R)$

$$||\Phi_\mu(f)||_{\mathcal{M}^p} = \overline{\lim_{T\to\infty}} \left(\frac{1}{2T}\int_{-T}^{T} |\Phi_\mu(f)|^p\right)^{\frac{1}{p}} \tag{5.1}$$

$$= \overline{\lim_{T\to\infty}} \left(\frac{1}{2T}\int_R |\Phi_\mu(\chi_{[-T,T]} f)|^p\right)^{\frac{1}{p}}$$

$$\leq ||\Phi_\mu||_{L^p} \cdot ||f||_{\mathcal{M}^p} .$$

Conversely, if for any $\varepsilon > 0$, there exists an $f \in L^p(R)$ such that

$$||\Phi_\mu(f)||_{L^p} \geq ||\Phi_\mu||_{L^p} \cdot ||f||_{L^p} - \varepsilon$$

we can construct an $\tilde{f}$ in $\mathcal{M}^p_r(R)$ such that

$$||\Phi_\mu(\tilde{f})||_{\mathcal{M}^p} \geq ||\Phi_\mu||_{L^p} \cdot ||\tilde{f}||_{\mathcal{M}^p} - \varepsilon.$$

This fact and (5.1) imply that

$$||\Phi_\mu||_{M^p} = ||\Phi_\mu||_{L^p} \qquad \text{for} \qquad \mu \in M_1.$$

It is known that for $\mu \in M$ and for $p = 1$, $||\Phi_\mu||_{L^1} = ||\mu||$; for $p = 2$, $||\Phi_\mu||_{L^2} = ||\hat{\mu}||_\infty$ where $\hat{\mu}$ is the Fourier-Stieltjes transformation of μ; and for $1 < p < \infty$, $p \neq 2$,

$$||\hat{\mu}||_\infty \leq ||\Phi_\mu||_{L^p} \leq ||\mu||.$$

If $\mu \in M$, then there exists a sequence of $\{\mu_n\}$ in M_1 which converges to μ, hence $\{\Phi_{\mu_n}\}$ is a Cauchy sequence of operators in $M^p(R)$ and converges to the operator Φ_μ.

<u>Theorem 5.2</u>. For $1 < p < \infty$ and for $\mu \in M$, the convolution operator $\Phi_\mu: M^p_r(R) \to M^p_r(R)$ is well defined and $||\Phi_\mu||_{M^p} = ||\Phi_\mu||_{L^p}$.

Let $g \in V^2(R)$ and let g be the inverse Fourier transformation of g, then the function defined by $f(t) = i\sqrt{2\pi}\, t\cdot g(t)$ is in $M^2(R)$ and $W(f) = g$ (see (4.1)). For each $\mu \in M_1$, we have

$$\Phi_\mu(f)(t) = -i\sqrt{2\pi}\, t\cdot\Phi_\mu(g)(t) = -i\sqrt{2\pi}\, t\cdot(\mu * g(t)) = -i\sqrt{2\pi}\, t\cdot(\hat{\mu}\cdot g)(t).$$

This implies that for $\mu \in M_1$,

$$W(\mu * f) = W(\Phi_\mu f) = \hat{\mu}\cdot g = \hat{\mu}\circ Wf.$$

By using a limit argument, we can show that:

<u>Theorem 5.4</u>. For each $\mu \in M$, the convolution operator $\Phi_\mu: M^2_r(R) \to M^2_r(R)$ defines the multiplier operator $\Psi_{\hat{\mu}}: V^2_r(R) \to V^2_r(R)$

$(V_r^2(R) = W(M_r^2(R)))$ which satisfies

$$W(\Phi_\mu(f)) = \Psi_{\hat{\mu}}(W(f))$$

Moreover, $||\Phi_\mu|| = ||\hat{\mu}||_\infty$ and

$$C^{-1}||\hat{\mu}||_\infty \leq ||\Psi_{\hat{\mu}}||_{V^2} \leq ||\hat{\mu}||_\infty , \qquad \mu \in M$$

where $C = ||W||\cdot||W^{-1}||$.

References

1. J. Bass, Stationary functions and their applications to the theory of turbulence, I. II, J. Math. Anal. & Appl. 47 (1974), 354-399, 458-503.

2. J. Bertrandias, Espaces de fonctions bornes et continues en moyenne asymptotique d'ordre p, Bull. Soc. Math. France, (1966), Memoire 5.

3. H. Bohr and E. Følner, On some type of functional spaces, a contribution to the theory of almost periodic functions, Acta. Math. 76 (1944), 31-155.

4. J. Doob, Time series and harmonic analysis, Proc. Berkeley Symposium on Mathematical Statistics and Probability, Univ. of Calif, Press, Berkeley, Calif., 1949, 303-343.

5. G. Hardy and J. Littlewood, some properties of fractional integrals I, Math. Zeit. 27 (1928), 565-606.

6. K. Lau, On the Banach spaces of functions with bounded upper means, Pacific J. Math (to appear).

7. _____, The class of convolution operators on the Marcinkiewicz spaces (to appear).

8. J. Lee and K. Lau, On generalized harmonic analysis, Tran. Amer. Math. Soc. 259 (1980), 75-97.

9. W. Luxemburg and A, Zaanen, Notes on Banach function spaces I, Indag. Math. 25 (1963), 135-147.

10. G. Mackey, Ergodic theory and its significance for statistical mechanic and probability theory, Advances in Math. 12 (1974), 178-268.

11. J. Marcinkiewicz, une remarque sur les espaces de M. Besicovitch, C. R. Acad. Sci. Paris, 208 (1939), 157-159.

12. P. Masani, Wiener's contribution to generalized harmonic analysis, prediction theory and filter theory, Bull. Amer. Math. Soc. 72 (1966), No. 1, 69-125.

13. _____, On helixes in Banach spaces, Sankhyā: Indian J. Stat. 38 (1976), 1-27.

14. _____, An outline of vector graph and conditional Banach spaces, Linear spaces and approximation, Edited by P. Butzer and B. Sz-Nagy, Birhauser Verlag Basel, 1978, 72-89.

15. _____, Comentary on the memoire on generalized harmonic analysis [30a], Norbert Wiener: Collected Works, Vol. II, Edited by Masani, (to appear).

16. R. Nelson, The space of functions of finite upper variations, Tran. Amer. Math. Soc. (to appear).

17. _____, Pointwise evaluation of Bochner integrals in Marcinkiewicz spaces (to appear).

18. N. Wiener, Generalized harmonic analysis, Acta. Math. 55 (1930).

19. _____, The Fourier integrals and certain of its applications, Dover, New York, 1959.

University of Pittsburgh
Pittsburgh, PA 15260

Contemporary Mathematics
Volume 2
1980

APPLIED FUNCTORIAL SEMANTICS, III: CHARACTERIZING BANACH CONJUGATE SPACES

F. E. J. Linton[1]

ABSTRACT. Using arguments suggested by functorial semantics, but calling only upon elementary functional analysis, we identify Banach conjugate spaces as the Eilenberg-Moore algebras for a certain monad on Banach spaces. Further investigation of certain topological and linear structures associated with such algebras leads to a deeper understanding of various Banach conjugate space characterization theorems of Dixmier, Waelbroeck, and Ng.

PREFACE. The dualization functor on Banach spaces, the contravariant functor which assigns to each space X (resp., to each map $f\colon X \longrightarrow Y$) its conjugate X^* (resp., $f^*\colon Y^* \longrightarrow X^*$), has been known for some time now to be monadic. The first ad hoc proof appeared in [12] and [13]; a coherent account of their disorganized and in part rather sketchy arguments appears in Semadeni's survey [20]. In his 1974 Warsaw dissertation [21], which may be more readily accessible as the IMPAN preprint [22], T. Świrszcz presents a perceptibly improved argument founded on suitable extensions of ideas in [8] and [10].

The present note, a long-overdue recension of the inaccessible original notes [12] and [13], gives yet another proof, one requiring minimal familiarity with functorial semantics; and goes on to expose the intimate relations this monadicity result bears to well known characterization theorems of Dixmier [2], Waelbroeck [23], [24], and Ng [18] for Banach conjugate spaces.

In outline: after setting the stage, we state the main theorem; we prove it; and we analyse it until the cited characterization theorems fall out with an easy shake.

This work is the outgrowth of a line of research first embarked on in 1969-70, in response to a question of John Isbell, while the author, on leave from his home university and supported in part by Canadian N. R. C. Grant no. A 7565, was a Killam Senior Research Fellow at Dalhousie University, Halifax, Nova Scotia. The author's gratitude to Arnold Tingley and Bill Lawvere, who

---

1980 Mathematics Subject Classification. 46B10, 18C20, 54D30, 46M99.
[1]During the final preparation of this work, the author was supported by National Science Foundation Grant no. MCS 80-06177.
Copyright © 1980, American Mathematical Society

made that seminal year possible, knows no bounds. Portions of this work, at various stages in its development, were presented (over the years following the appearance of [12] and [13]) to the American Mathematical Society at St. Louis in April, 1972 (see [14]), and to seminars, colloquia, and symposia at Oberwolfach, the Banach Center (Warsaw), and the Universities of Louvain, Linz, Bremen, Hagen, Konstanz, and Carleton (Ottawa): to all my hosts and auditors I express my thanks for their welcome criticism, support, and interest. I am grateful as well for the financial support provided by the University of North Carolina (Chapel Hill) and, year in and year out, by Wesleyan University.

Further thanks are due to George Reynolds, who first made me aware of [18]; and, especially, to Bill Graves and Barbara Janson: what lies before you owes its existence as much to the support and kind encouragement provided by the one as to the firm deadline prescribed by the other.

Finally -- and filially, as it were -- I acknowledge my incalculable indebtedness to three giants upon whose different shoulders I have long ridden -- and to whom this work is dedicated -- three minds untimely stilled by cancer: S. Banach, B.J. Pettis, and M. Linton.

1. THE MONADICITY THEOREM. We write Ban for the category of real (or complex) Banach spaces. This is at once:

(i) an ordinary category, if we take as morphisms $f: X \longrightarrow Y$ the linear transformations f (from X to Y) of bound $\|f\| \leq 1$, i.e., the contractive linear transformations;

(ii) a concrete category, if we exploit the "underlying set" functor $d: \mathrm{Ban} \longrightarrow \mathrm{Sets}$ defined by the passage from a Banach space X (resp., a contractive linear transformation $f: X \longrightarrow Y$) to its unit disc $d(X) = X_1 = \{x \in X \,/\, \|x\| \leq 1\}$ (resp., its restriction $f|_{d(X)}$, constrained to a function $d(f): d(X) \longrightarrow d(Y)$);

(iii) a closed category (cf. [4]), if, exploiting the observation that the hom sets of the ordinary category Ban are just the unit discs of the Banach spaces $\mathrm{Ban}(X, Y)$ of all bounded linear transformations $f: X \longrightarrow Y$, we lift the set-valued hom functor of Ban up, over the unit disc functor d, to an internal, i.e., Ban-valued, hom functor $\mathrm{Ban}^{op} \times \mathrm{Ban} \longrightarrow \mathrm{Ban}$ taking precisely the Banach spaces $\mathrm{Ban}(X, Y)$ as values; and

(iv) a Ban-category (that is, a $\mathcal{V}$-category in the sense of [4], where $\mathcal{V} = \mathrm{Ban}$).

Moreover, the closed category Ban is symmetric, the symmetry being given by the familiar isometric isomorphisms

$$t_{XYZ}: \mathrm{Ban}(X, \mathrm{Ban}(Y, Z)) \xrightarrow{\cong} \mathrm{Ban}(Y, \mathrm{Ban}(X, Z))$$

assigning to a bounded linear transformation $f: X \longrightarrow \mathrm{Ban}(Y, Z)$ the bounded linear transformation $t_{XYZ}(f) = f^t: Y \longrightarrow \mathrm{Ban}(X, Z)$, <u>transpose</u> of f, for which the defining equation is

$$\{ f^t(y) \}(x) = \{ f(x) \}(y) \quad (y \in Y, \ x \in X) .$$

The verifications of the formula $f^{tt} = f$ and of the naturality of t_{XYZ} in the variables X, Y, and Z, are routine.

In particular, each contravariant functor $\overline{D_Z} = \mathrm{Ban}(-, Z): \mathrm{Ban} \longrightarrow \mathrm{Ban}$ is its own right adjoint; equivalently, replacing the contravariant functor $\overline{D_Z}$ by each of the covariant functors

$$D_Z: \mathrm{Ban}^{op} \longrightarrow \mathrm{Ban} , \quad D_Z^{\ op}: \mathrm{Ban} \longrightarrow \mathrm{Ban}^{op}$$

with which it is conventionally identified (cf. pp. 33-34 and 86-87 of [16]), the symmetry of Ban results in an adjunction situation

$$(1.1) \qquad t: D_Z^{\ op} \dashv D_Z \ : \ (\mathrm{Ban}^{op}, \ \mathrm{Ban})$$

in which $D_Z^{\ op}$ is left adjoint (coadjoint) to D_Z.

Our sole concern, as regards these adjunction situations, is with the one in which the dualizing object Z is the scalar field itself. And in that case, we write simply D and D^{op} for D_Z and $D_Z^{\ op}$, and adopt the more customary conjugate space notations $X^* = D(X)$, $f^* = D(f): Y^* \longrightarrow X^*$ for the conjugate space / conjugate transformation values of D at an object or a map.

We shall need the upshot of the standard construction of the monad, or triple, $\nabla(t)$, <u>generated</u> in Ban by the adjunction situation (1.1). Consulting [5: Prop. 2.1] or [16: pp. 81, 134], this is (for Z the scalar field) the <u>double</u> <u>dualization</u> monad $\nabla(t) = ** = (DD^{op}, \ i, \ \mu)$, where the natural transformation $i: \mathrm{id}_{\mathrm{Ban}} \longrightarrow DD^{op}$ has as components $i_X: X \longrightarrow DD^{op}X$ the familiar <u>evaluation</u> maps $i_X = (\mathrm{id}_{X^*})^t: X \longrightarrow X^{**} = DD^{op}X$ (defined, for $x \in X$ and $\varphi \in X^*$, by $\{ i_X(x) \}(\varphi) = \varphi(x)$), while $\mu: DD^{op}DD^{op} \longrightarrow DD^{op}$ is the natural transformation given by $\mu_X = (i_{X^*})^*: X^{****} \longrightarrow X^{**}$.

Similarly, we shall need the upshot of the standard construction of the semantical comparison functor $\Phi: \mathrm{Ban}^{op} \longrightarrow (\mathrm{Ban})^{**}$ between Ban^{op} and the Eilenberg-Moore category $(\mathrm{Ban})^{**}$ of $**$-algebras in Ban. Consulting [5: Th. 2.2] or [16: pp. 136-139], $(\mathrm{Ban})^{**}$ is the category whose objects (the $**$-algebras) are pairs (X, ξ) consisting of a Banach space X (the underlying Banach space of (X, ξ)) and a contractive linear transformation $\xi: X^{**} \longrightarrow X$ (the structure map of (X, ξ)) rendering commutative the diagrams

$$\begin{array}{ccc} X & \xrightarrow{\ i_X\ } & X^{**} \\ & \underset{\mathrm{id}_X}{\searrow} & \downarrow \xi \\ & & X \end{array} , \qquad \begin{array}{ccc} X^{****} & \xrightarrow{\ \xi^{**}\ } & X^{**} \\ \mu_X \downarrow & & \downarrow \xi \\ X^{**} & \xrightarrow[\ \xi\ ]{} & X \end{array} .$$

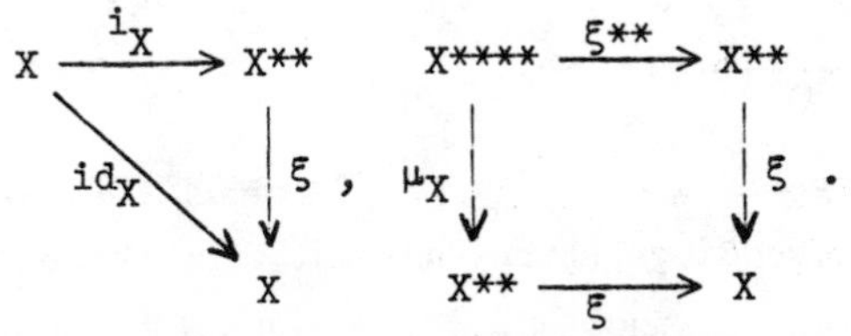

Like Ban , $(\mathrm{Ban})^{**}$ is both an ordinary category and a Ban-category, as follows. A $**$-homomorphism from one $**$-algebra (X, ξ) to another (Y, η) is any bounded linear transformation $f\colon X \longrightarrow Y$ making the square

$$\begin{array}{ccc} X^{**} & \xrightarrow{f^{**}} & Y^{**} \\ \xi \downarrow & & \downarrow \eta \\ X & \xrightarrow[f]{} & Y \end{array}$$

commute, or -- equivalently -- lying in the kernel of the contractive linear transformation $\mathrm{Ban}(X, Y) \longrightarrow \mathrm{Ban}(X^{**}, Y)$ sending f to $\frac{1}{2}(\eta\circ f^{**} - f\circ\xi)$. The Ban-category $(\mathrm{Ban})^{**}$ uses these kernels as the values $(\mathrm{Ban})^{**}((X, \xi),(Y,\eta))$ of its Ban-valued hom functor, while the ordinary category $(\mathrm{Ban})^{**}$ uses their unit discs as the values of its set-valued hom functor (thus the $(\mathrm{Ban})^{**}$-morphisms from (X, ξ) to (Y, η) are the contractive $**$-homomorphisms between them).

The functor Φ is now quite easy to describe: writing $\varphi_A = (i_A)^*\colon A^{***} \longrightarrow A^*$, one observes that (A^*, φ_A) is a $**$-algebra and that, whatever the bounded linear transformation $f\colon B \longrightarrow A$, its conjugate $f^*\colon A^* \longrightarrow B^*$ is a $**$-homomorphism from (A^*, φ_A) to (B^*, φ_B) -- and one simply sets $\Phi(A) = (A^*, \varphi_A)$, $\Phi(f) = f^*$. For universal properties of Φ, both as ordinary functor and as Ban-functor, see [5], [16: §VI.3], and [11: §2].

One last ingredient remains to be stirred in. To make the statement of our main theorem a theorem at all, we need (essentially for the Hahn-Banach theorem) the Axiom of Choice. But then:

THEOREM. The semantical comparison functor $\Phi\colon \mathrm{Ban}^{\mathrm{op}} \longrightarrow (\mathrm{Ban})^{**}$ is an equivalence of categories (both at the ordinary level and at the level of Ban-categories).

2. THE PROOF. We must show two things:

(2.1) Φ is fully faithful; and

(2.2) every $**$-algebra (X, ξ) is isomorphic with a value (A^*, φ_A) of Φ.

Observe first that each square

$$(2.3)\qquad \begin{array}{ccc} A & \xrightarrow{f} & B \\ i_A \downarrow & & \downarrow i_B \\ A^{**} & \xrightarrow[f^{**}]{} & B^{**} \end{array} \qquad (f\colon A \longrightarrow B \text{ a bounded linear transformation})$$

commutes (naturality of $i = (i_A)_A\colon \mathrm{id}_{\mathrm{Ban}} \longrightarrow DD^{\mathrm{op}}$), and that the maps i_A, i_B are isometric embeddings (Hahn-Banach). The consequent norm inequalities

$$\|f\| = \|i_B\circ f\| = \|f^{**}\circ i_A\| \leq \|f^{**}\| \leq \|f^*\| \leq \|f\|$$

make an isometric embedding of the linear transformation

$$f \longmapsto \Phi(f) : Ban^{op}(B, A) = Ban(A, B) \longrightarrow (Ban)^{**}(\Phi(B), \Phi(A)),$$

whence Φ is faithful.

To finish the proof of (2.1), we rely on a corollary of the following useful but little-known instance of the Hahn-Banach theorem.

LEMMA 1 ([12], Lemma 1.1). When $f: A \longrightarrow B$ is an isometric embedding, the commutative square (2.3) is a pullback diagram in Ban.

PROOF. Suppose $F \in A^{**}$ and $b \in B$ satisfy $f^{**}(F) = i_B(b)$. Then each continuous functional h on B, vanishing on the closed subspace $f(A)$, must vanish at b as well:

$$h(b) = \{i_B(b)\}(h) = \{f^{**}(F)\}(h) = \{F \cdot f^* \}(h) = F(f^*(h)) = F(h \cdot f) = F(0) = 0.$$

Consequently (Hahn-Banach) $b \in f(A)$, i.e., there is $a \in A$ with $f(a) = b$. But for this a, already unique because f was one to one, we have

$$f^{**}(F) = i_B(b) = i_B(f(a)) = j^{**}(i_A(a)),$$

which completes the proof of the lemma.

COROLLARY. Whatever the Banach space X, $\quad X \xrightarrow{\; i_X \;} X^{**} \underset{(i_X)^{**}}{\overset{i_{X^{**}}}{\rightrightarrows}} X^{****}$ is an equalizer diagram.

PROOF. We may apply the lemma to $A = X$, $B = X^{**}$, $f = i_X$ because the Hahn-Banach theorem assures that $f = i_X$ is an isometric embedding. In the resulting pullback diagram (draw it) the two maps from X are the same, and the corollary follows.

Now, to finish the proof that Φ is fully faithful, let $g: \Phi(B) \longrightarrow \Phi(A)$ be given. Because g is a *-homomorphism, the square

(2.4)

$$
\begin{array}{ccc}
B^{***} & \xrightarrow{\;g^{**}\;} & A^{***} \\
{\scriptstyle (i_B)^* \,=\, \varphi_B}\big\downarrow & & \big\downarrow{\scriptstyle \varphi_A \,=\, (i_A)^*} \\
B^* & \xrightarrow[g]{} & A^*
\end{array}
$$

commutes. Consider the (partial) map of equalizer diagrams

$$
\begin{array}{ccccc}
A & \xrightarrow{\;i_A\;} & A^{**} \underset{(i_A)^{**}}{\overset{i_{A^{**}}}{\rightrightarrows}} & A^{****} \\
{\scriptstyle \exists ? f}\big\downarrow & & {\scriptstyle g^*}\big\downarrow & & \big\downarrow{\scriptstyle g^{***}} \\
B & \xrightarrow[i_B]{} & B^{**} \underset{(i_B)^{**}}{\overset{i_{B^{**}}}{\rightrightarrows}} & B^{****} \; .
\end{array}
$$

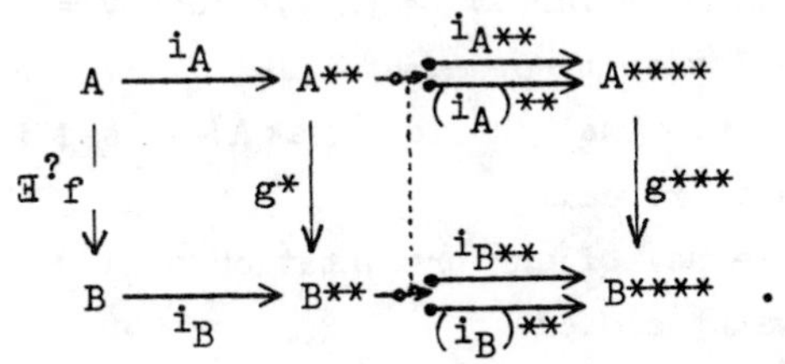

The right hand square commutes whether the DPDT switch is up (naturality of i) or down (conjugate of the commutative square (2.4)). Hence $g^* \cdot i_A$ factors uniquely, where indicated by the broken arrow, through the equalizer i_B of the lower row. If $f: A \longrightarrow B$ is that factorization, we have $i_B f = g^* i_A$ (whence f is necessarily bounded), and g^* and f^{**}, both (weak A^*, weak B^*)-continuous from A^{**} to B^{**}, agree on the (weak A^*)-dense subspace $i_A(A)$ (proof: $f^{**} i_A = i_B f = g^* i_A$), hence are equal, $g^* = f^{**}$. But then $g = f^*$ (by the faithfulness already established), and Φ is fully faithful.

We are now free to turn to the proof of (2.2). By a <u>presentation</u> of a $**$-algebra (X, ξ) as a quotient of a value $\Phi(A)$ of Φ, we shall mean a contractive $**$-homomorphism $p: \Phi(A) = (A^*, \varphi_A) \longrightarrow (X, \xi)$ which is at the same time a quotient map $A^* \longrightarrow X$ in Ban. For example, the algebra structure $\xi: X^{**} \longrightarrow X$ itself is a presentation of (X, ξ) as a quotient of $\Phi(X^*)$.

Fixing for the duration a $**$-algebra (X, ξ) and a presentation p of (X, ξ) as a quotient of $\Phi(A)$, let K (with inclusion $j: K \longrightarrow A^*$) be the kernel of the contractive quotient map $p: A^* \longrightarrow X$, let $v: V \longrightarrow A$ be the inclusion in A of the polar $V = K^\perp$ of K, that is, the kernel of the transpose $j^t = j^* \circ i_A : A \longrightarrow K^*$ of j, and let $q: A \longrightarrow A/V$ be the cokernel of v. Then, comparing the short exact sequence $K \xrightarrow[j]{} A^* \xrightarrow[p]{} X$ with the conjugate of the short exact sequence $V \xrightarrow[v]{} A \xrightarrow[q]{} A/V$, we obtain a diagram of short exact sequences

$$\begin{array}{ccccc} K & \xrightarrow[j]{} & A^* & \xrightarrow[p]{} & X \\ \alpha \downarrow & & \downarrow = & & \downarrow \beta \\ (A/V)^* & \xrightarrow[q^*]{} & A^* & \xrightarrow[v^*]{} & V^* \end{array} ,$$

the maps α and β being the unique maps to the kernel of v^* and from the cokernel of j reflecting the fact that $v^* \circ j = 0$ (proof: $v^* \circ j = (j^t \circ v)^t$).

Note that β is then a $**$-homomorphism from (X, ξ) to $\Phi(V)$: one proves $\beta \xi = \varphi_V \beta^{**}$ by simply canceling the epimorphism p^{**} from the end result of the calculation

$$\beta \xi p^{**} = \beta p \varphi_A = v^* \varphi_A = (i_A v)^* = (v^{**} i_V)^* = \varphi_V v^{***} = \varphi_V \beta^{**} p^{**} .$$

Moreover, necessary and sufficient for β to be an isometric isomorphism is that α be one. For α to be one, however, since q^* is essentially the inclusion in A^* of the polar $V^\perp \cong (A/V)^*$ of $V = K^\perp$, it is necessary and sufficient (by the bipolar theorem, cf. [19: IV.1.5 (p. 126)]) that K be (weak A)-closed in A^*. That K is (weak A)-closed it now becomes the task of the following lemma to reveal.

LEMMA 2. The kernel of any presentation of a $**$-algebra as a quotient of a value of Φ is weak*-closed.

PROOF. Retaining our earlier notation, let us show that the $**$-algebra structure φ_A on A^* carries $K^{**} \xrightarrow[j^{**}]{} A^{***}$ into $K \xrightarrow[j]{} A^*$ and the unit disc dK^{**} of K^{**} onto the unit disc dK of K. Indeed,

$$p\,\varphi_A\, j^{**} = \xi\, p^{**}\, j^{**} = \xi\circ(pj)^{**} = \xi\circ 0 = 0\ ,$$

i.e., $\varphi_A \circ j^{**}$ is annihilated by p, hence factors uniquely -- by, say, $k\colon K^{**} \longrightarrow K$ -- through j, the kernel of p, as indicated in the diagram

$$\begin{array}{ccc} K^{**} & \xrightarrow{\ j^{**}\ } & A^{***} \\ {\scriptstyle k}\big\downarrow & & \big\downarrow {\scriptstyle \varphi_A \,=\, (i_A)^{*}} \\ K & \xrightarrow[\ j\ ]{} & A^{*} \end{array} \quad .$$

But now, for any $x \in K$, $k(i_K(x)) = x$, as is seen by canceling the monomorphism j from the end result of the calculation

$$j\,k\,i_K = \varphi_A\, j^{**}\, i_K = \varphi_A\, i_{A*}\, j = j = j\circ \mathrm{id}_K\ ;$$

and $\|k\|$ is clearly $\leqq 1$. In fact, (K, k) is a ⁂-algebra, as anyone aware that monadic functors create limits (cf. [9: §6] or [17: 3.1.17-.19]) will have recognized, but this fact is not exploited in what follows. Instead, we observe that the unit disc of K, image of the (weak K^*)-compact unit disc of K^{**} under the (weak K^*, weak A)-continuous map $\varphi_A\circ j^{**}\colon K^{**} \longrightarrow A^*$, is a (weak A)-compact subset of the (weak A)-separated space A^*, hence is (weak A)-closed in A^*, so that, by the baby Krein-Šmulian theorem (in the form of the corollary to IV.6.4 in [19] -- but see also [3: V.5.8] or, as is pointed out in [15: p. 313, Remark 2], even [1: p. 121, Lemma 3]), K itself is (weak A)-closed in A^*, as was to be shown.

This completes the proof of the theorem.

3. BANACH PRECONJUGATES. It is a simple consequence of the monadicity theorem of §1 that a Banach space is (isometrically isomorphic with) a conjugate space if and only if it admits the structure of a ⁂-algebra. In fact, much more is true: the ⁂-algebra structures a given space X admits completely classify the ways in which X can arise as a conjugate space, and certain equivalence classes of these ⁂-algebra structures classify the isomorphism types of the Banach spaces whose conjugate X can be, as we shall now see.

We begin with some words about the ways in which a fixed Banach space X can arise as a conjugate space. By a (Banach) preconjugate for X we mean simply the data required for a representation of X as a Banach conjugate space, that is, a pair (A, a) with A a Banach space and $a\colon X \longrightarrow A^*$ an isometric isomorphism. Two preconjugates for X, (A, a) and (B, b), are <u>equivalent</u> if the representations they provide are indistinguishable, that is, if there is an isometric isomorphism $f\colon A \longrightarrow B$ for which $f^*\circ b = a$. A preconjugate (A, a) for X is in <u>canonical form</u> if A is literally a subspace of X^*, and if a is actually the transpose, $a = v^t$, of the literal inclusion function $v\colon A \xrightarrow{\ \subset\ } X^*$ of A in X^*.

Note that each preconjugate (A, a) for X is equivalent to one in canonical form. Indeed, let B be the image of the isometric embedding of A into X^* provided by the transpose $a^t = a^* \circ i_A$ of the isometric isomorphism $a: X \longrightarrow A^*$, and, writing $v: B \longrightarrow X^*$ for the inclusion, observe that the obvious isomorphism $f: A \longrightarrow B$ induced by a^t serves to mediate equivalence of (A, a) with (B, v^t). Note also that if equivalent preconjugates (A, a) and (B, b) for X are already in canonical form, the fact that a^t and b^t are literal inclusions and the relation $b^t \circ f = a^t$ (obtained by transposing $f^* \circ b = a$) force the subspaces A and B of X^* to coincide. Thus, a complete system of mutually inequivalent representatives for the equivalence classes of Banach preconjugates for X is provided by the set $\mathrm{Pre}_c(X)$ of preconjugates in canonical form.

Now observe that, if (A, a) is a preconjugate for X, the map

$$(3.1) \qquad \xi_{(A, a)} = a^{-1} \circ \varphi_A \circ a^{**} : X^{**} \longrightarrow A^{***} \longrightarrow A^* \longrightarrow X$$

is the unique $\ast\ast$-algebra structure $\xi: X^{**} \longrightarrow X$ on X for which a is a $\ast\ast$-homomorphism $(X, \xi) \longrightarrow \Phi(A)$. It follows that the $\ast\ast$-algebra structures $\xi_{(A, a)}$, $\xi_{(B, b)}$ on X, associated in this way with equivalent preconjugates (A, a), (B, b) for X, must coincide: $\xi_{(A, a)} = \xi_{(B, b)}$. Conversely, if the $\ast\ast$-algebra structures on X so associated with two preconjugates for X coincide, $\xi_{(A, a)} = \xi_{(B, b)} = \xi$, say, then the composition

$$a \circ b^{-1}: \Phi(B) \longrightarrow (X, \xi) \longrightarrow \Phi(A)$$

is an isomorphism in $(\mathrm{Ban})^{\ast\ast}$ from $\Phi(B)$ to $\Phi(A)$, hence is of the form $\Phi(f)$ for a unique isometric isomorphism $f: A \longrightarrow B$, by the monadicity theorem; but then, from the relation $f^* = a \circ b^{-1}$, it follows that $f^* \circ b = a$, whence (A, a) and (B, b) were equivalent. With one last remark, namely, that (2.2) guarantees each $\ast\ast$-algebra structure on X to be of the form $\xi_{(A, a)}$ for some preconjugate (A, a) for X (the proof even exhibited (A, a) in canonical form), we have established part (i) of the following proposition.

PROPOSITION 1. (i) The passage $(A, a) \mapsto \xi_{(A, a)}$ from a preconjugate (A, a) for the Banach space X to the $\ast\ast$-algebra structure $\xi_{(A, a)}$ on X, defined by (3.1), puts the family $\mathrm{Pre}(X)$ of equivalence classes of Banach preconjugates for X -- for which the set $\mathrm{Pre}_c(X)$ of preconjugates in canonical form is a complete irredundant system of representatives -- in bijective correspondence with the set $\mathrm{Alg}(X)$ of $\ast\ast$-algebra structures on X.

(ii) Preconjugates (A, a) and (B, b) for X, not necessarily equivalent, nonetheless have isometrically isomorphic spaces A and B if and only if the associated $\ast\ast$-algebra structures $\xi_{(A, a)}$ and $\xi_{(B, b)}$ are spatially linked by a linear operator on X which is both an isometric isomorphism of X onto itself and a $\ast\ast$-homomorphism from $(X, \xi_{(A, a)})$ to $(X, \xi_{(B, b)})$.

PROOF. It remains to establish (ii). But, by the monadicity theorem, passing from $f \in \mathrm{Ban}(B, A)$ to $b^{-1} \circ \Phi(f) \circ a \in (\mathrm{Ban})^{**}((X, \xi_{(A,a)}), (X, \xi_{(B,b)}))$ induces a bijective correspondence between the isometric isomorphisms from B to A and the $**$-isomorphisms between $(X, \xi_{(A,a)})$ and $(X, \xi_{(B,b)})$.

4. DIXMIER COMPLEMENTS. A subspace $K \subset X^{**}$ is a <u>Dixmier complement</u> for the Banach space X if it is (weak X^*)-closed and algebraically complementary to $i_X(X) \subset X^{**}$, and if there prevail the norm inequalities

$$\|x\| \leqq \|i_X(x) + k\| \qquad (\text{all } x \in X, \text{ all } k \in K). \tag{4.1}$$

These are the subspaces of X^{**} whose existence served Dixmier as a necessary and sufficient condition for X to be a Banach conjugate space (Th. 18 in [2]).

PROPOSITION 2. Let X be a Banach space. By simple restriction, the passage $\xi \mapsto \ker(\xi)$ from an arbitrary linear transformation $\xi: X^{**} \longrightarrow X$ to its kernel $\ker(\xi) = \xi^{-1}(0) \subset X^{**}$ sets up bijective correspondences between:

(i) linear retractions along i_X and algebraic complements in X^{**} to $i_X(X)$,

(ii) contractive such retractions and such complements satisfying (4.1),

(iii) $\mathrm{Alg}(X)$ and Dixmier complements for X.

PROOF. (i) is completely standard (see [6: §29]) once it is explained that a function $\xi: X^{**} \longrightarrow X$ is being called a linear retraction along i_X iff it is a merely algebraic linear transformation satisfying $\xi \circ i_X = id_X$. The unique linear retraction ξ corresponding to -- i.e., with kernel -- a given algebraic complement K to $i_X(X)$ is defined, for $x \in X$ and $k \in K$, by $\xi(i_X(x) + k) = x$. It follows easily that $\|\xi\| \leqq 1$ iff (4.1) holds, which establishes (ii). As for (iii), recalling Lemma 2, which assures us that the kernel of a $**$-algebra structure on X must be (weak X^*)-closed, we see that we need only settle the converse, i.e., we must show that if the kernel of a contractive linear retraction is (weak X^*)-closed, then that retraction was a $**$-algebra structure on X. One of the algebra equations ($\xi \circ i_X = id_X$) being valid by hypothesis, it remains only to establish the other ($\xi\xi^{**} = \xi\mu_X$). And for this, writing K for $\ker(\xi)$ and $j: K \longrightarrow X^{**}$ for the inclusion, notice that, as $j^{**}(K^{**})$ and $(i_X)^{**}(X^{**})$ remain algebraically complementary in X^{****} when K and $i_X(X)$ are algebraically complementary in X^{**}, it will be enough to establish

$$\xi \circ \xi^{**} \circ (i_X)^{**} = \xi \circ \mu_X \circ (i_X)^{**} \quad \text{and} \quad \xi \circ \xi^{**} \circ j^{**} = \xi \circ \mu_X \circ j^{**}.$$

The first of these is trivial, since $\xi i_X = id_X$ and $\mu_X \circ (i_X)^{**} = id_{X^{**}}$. For the second, notice that the (weak X^*)-closedness of K makes the topology τ on X, quotient via ξ of the (weak X^*)-topology on X^{**}, separated; and of course ξ becomes (weak X^*, τ)-continuous. So, to establish equality of the two (weak K^*, τ)-continuous functions $\xi \circ \xi^{**} \circ j^{**}$ and $\xi \circ \mu_X \circ j^{**}$, it suffices to establish their equality on a convenient (weak K*)-dense subspace of K^{**}. What comes to mind is $i_K(K)$ -- and there they both vanish:

$$\xi \circ \xi^{**} \circ j^{**} \circ i_K = \xi \circ (\xi j)^{**} \circ i_K = \xi \circ 0 \circ i_K = 0,$$

$$\xi \circ u_X \circ j^{**} \circ i_K = \xi \circ \mu_X \circ i_{X^{**}} \circ j = \xi \circ id_{X^{**}} \circ j = \xi \circ j = 0.$$

This completes the proof.

5. COMPACT TOPOLOGIES ON THE UNIT DISC. Having now used for the second time the fact that the kernel of a **-algebra structure $\xi: X^{**} \longrightarrow X$ is (weak X*)-closed, it is time to point out that, in the (separated) topology τ, quotient via ξ of the (weak X*)-topology on X^{**}, the unit disc dX of X, image under ξ of the (weak X*)-compact unit disc of X^{**}, is compact. By Alaoglu, that's no surprise: certainly, at least when X is a conjugate space A* and $\xi = \varphi_A$ -- situation that, by the monadicity theorem, is essentially typical -- the disc dX is (weak A)-compact; and, clearly, the comparable, separated topologies τ and (weak A) coincide, at least, in their traces on that disc.

We write $\tau_\xi = \tau|_{dX}$ for that trace. The twin purposes of this section, now that we are in the presence of a function $\xi \mapsto \tau_\xi$ from the set Alg(X) to the set of compact Hausdorff topologies on the unit disc dX of X, are: (i) to show that this function is one to one; and (ii) to characterise its image. In fact, the image characterization comes first.

PROPOSITION 3. Let X be a Banach space, and let $\varkappa$ be a topology on the unit disc dX of X for which dX is $\varkappa$-compact. Then the following assertions (under each of which dX will be $\varkappa$-separated) are equivalent.

(i) There is a **-algebra structure ξ on X with $\varkappa = \tau_\xi$.

(ii) ([18], [23]) $\varkappa$ is the trace on dX of some separated locally convex topology τ on X.

(iii) $\varkappa$ is the trace on dX of some topological linear space topology τ on X for which there are enough τ-continuous functionals to separate points.

(iv) ([24: 3.1(a)]) The functionals on X, $\varkappa$-continuous on dX, separate points.

(v) ([2: Th. 14, 17]) There is a norm-closed subspace $V \subset X^*$, separating points of X, with $\varkappa$ = trace on dX of the (weak V)-topology on X.

(vi) dX is $\varkappa$-separated, $\varkappa$ is weaker than the trace on dX of the (weak X*)-topology on X, and the functions $x \mapsto \alpha x$ ($|\alpha| \leqq 1$) and $(x, y) \mapsto \frac{1}{2}(x+y)$ are continuous with respect to $\varkappa$ and the product topology on $dX \times dX$ obtained from $\varkappa$. (Compare [24: §3].)

PROOF. (i) ⇒ (ii) is trivial (using $\tau = \tau_\xi$), as are (ii) ⇒ (iii) and (iii) ⇒ (iv). (iv) ⇒ (v) is easy: let V be the set of all linear forms on X, $\varkappa$-continuous on dX. For $f \in V$, f(dX) is compact, so f is bounded, and $V \subset X^*$. Restriction to dX maps V isometrically onto an equationally defined, hence norm-closed, subspace of $C((dX, \varkappa))$, so V is norm-closed in X^*, separating points by (iv). The (weak V)-topology on X is therefore separated; and its trace on dX, being both weaker than $\varkappa$, hence compact, and separated, must therefore coincide with $\varkappa$.

(v) ⇒ (vi) being trivial again, we turn to (vi) ⇒ (i). We begin by endowing X and X^{**} with the uniformities associated with the (weak X^*)-topologies (a typical basic entourage being the set $\{(x, y) \,/\, y - x \in U\}$ of all pairs (x, y) whose differences $y - x$ lie in a given (weak X^*)-open neighborhood U of 0). Giving their unit discs the relative uniformities, the (weak X^*)-compact disc dX^{**} we may view as the completion of its dense uniform subspace dX, under the inclusion $d(i_X)\colon dX \longrightarrow dX^{**}$. Similarly, we view the compact Hausdorff space $(dX, \varkappa)$ as a complete, separated, uniform space, and we extend the (weak X^*, $\varkappa$)-uniformly continuous identity function on dX over $d(i_X)$ to a (weak X^*, $\varkappa$)-uniformly continuous function on the completion, $\xi_1\colon dX^{**} \longrightarrow dX$, which then satisfies the equation: $\xi_1 \circ d(i_X) = id_{dX}$. The following lemma will help us realize that this function ξ_1 is the restriction to the unit discs, $\xi_1 = d(\xi)$, of a (unique) $*$-algebra structure ξ on X.

LEMMA 3. Let A and B be two Banach spaces, and $f_1\colon dA \longrightarrow dB$ a function between their unit discs. Necessary and sufficient for f_1 to be the restriction, $f_1 = d(f)$, of a (necessarily unique, and necessarily contractive) linear transformation $f\colon A \longrightarrow B$ is that f_1 satisfy the equations

$$f_1(\alpha x) = \alpha \cdot f_1(x)\,, \qquad f_1(\tfrac{1}{2}(x+y)) = \tfrac{1}{2}(f_1(x) + f_1(y))$$

for all $x, y \in dA$ and all scalars α with $|\alpha| \leqq 1$.

REMARK. This lemma may be viewed as a reflection of one aspect of the functorial semantics of the unit disc functor $d\colon \mathrm{Ban} \longrightarrow \mathrm{Sets}$.

PROOF. Suppose f_1 preserves the indicated subconvex combinations. Given $a \in A$, let λ and μ be scalars with $\mu \geqq \lambda \geqq \max(1, \|a\|)$. Then $0 < \mu^{-1}\lambda \leqq 1$ and $\lambda^{-1}a$, $\mu^{-1}a$ both are in dA, and we have

$$\lambda \cdot f_1(\lambda^{-1}a) = \mu\mu^{-1}\lambda \cdot f_1(\lambda^{-1}a) = \mu \cdot f_1(\mu^{-1}\lambda\lambda^{-1}a) = \mu \cdot f_1(\mu a)\,.$$

It is therefore unambiguous to define a function $f\colon A \longrightarrow B$ by the formula:

given $a \in A$, choose $\lambda \geqq \max(1, \|a\|)$, and set $f(a) = \lambda \cdot f_1(\lambda^{-1}a)$.

For $x \in dA$, then, we see $f(x) = f_1(x)$ by choosing $\lambda = 1$. Given a and $b \in A$, we choose $\lambda \geqq \|a\| + \|b\| + 1$, and calculate

$$\begin{aligned} f(a+b) &= 2\lambda \cdot f_1((2\lambda)^{-1}(a+b)) = 2\lambda \cdot f_1(\tfrac{1}{2}(\lambda^{-1}a + \lambda^{-1}b)) \\ &= 2\lambda \cdot \tfrac{1}{2}(f_1(\lambda^{-1}a) + f_1(\lambda^{-1}b)) = \lambda \cdot f_1(\lambda^{-1}a) + \lambda \cdot f_1(\lambda^{-1}b) \\ &= f(a) + f(b)\,. \end{aligned}$$

Similarly, given $x \in A$ and a scalar α, we choose large enough λ and see:

$$f(\alpha x) = \lambda^2 \cdot f_1(\lambda^{-2}\alpha x) = \lambda^2 \cdot f_1(\lambda^{-1}\alpha\lambda^{-1}x)$$
$$= \lambda^2\lambda^{-1}\alpha \cdot f_1(\lambda^{-1}x) = \alpha\cdot\lambda\cdot f_1(\lambda^{-1}x) = \alpha\cdot f(x) .$$

This, leaving the necessity (and the parenthetical remarks) to the reader, completes the proof of the lemma.

Returning to the proof of the proposition, observe that, whatever the scalars α, β, with $|\alpha| + |\beta| \leqq 1$, both $d(i_X)\colon dX \longrightarrow dX^{**}$ and id_{dX} preserve the subconvex combination operations $(x, y) \mapsto \alpha x + \beta y$. Because these are jointly (weak X^*)-continuous and $dX \times dX$ is (weak X^*)-dense in $dX^{**} \times dX^{**}$, the (weak X^*, $\varkappa$)-continuous function ξ_1 will preserve whichever of them are $\varkappa$-continuous on dX, viz., $(x, y) \mapsto \frac{1}{2}(x+y)$ and $x \mapsto \alpha x$.

Let $\xi\colon X^{**} \longrightarrow X$ be the contractive linear transformation extending ξ_1 provided by the lemma. The relation $\xi\circ i_X = id_X$ follows from $\xi_1\circ d(i_X) = id_{dX}$. Thus ξ is a contractive linear retraction along i_X, and to learn it is a $**$-algebra structure, it suffices (by Proposition 2) to prove that $\ker(\xi)$ is (weak X^*)-closed. But, in fact, $\ker(\xi)$ is a norm-closed subspace of X^{**}, and its unit disc, being the inverse image of the $\varkappa$-closed point 0 under ξ_1, is (weak X^*)-closed in dX^{**}, and hence also in X^{**}, and we appeal to the Krein-Šmulian theorem.

Finally, since both $\varkappa$ and τ_ξ are compact Hausdorff topologies, and since they are comparable ($\varkappa$ being obviously weaker, since ξ_1 is (weak X^*, $\varkappa$)-continuous), they coincide, which closes the circle and ends the proof.

Writing $\mathrm{Comp}(X)$ for the set of compact topologies, on the unit disc dX of the Banach space X, that satisfy the equivalent conditions of Proposition 3, it is high time to see that the surjection $\xi \mapsto \tau_\xi\colon \mathrm{Alg}(X) \longrightarrow \mathrm{Comp}(X)$ is one to one. To see it, one observes that the polar in X^* of $\ker(\xi) \subset X^{**}$ is the space of functionals τ_ξ-continuous on dX. The following lemma records that this observation is more generally valid.

LEMMA 4. Let X be a Banach space; let $\xi\colon X^{**} \longrightarrow X$ be a contractive linear retraction along $i_X\colon X \longrightarrow X^{**}$; let $K = \ker(\xi)$, with $j\colon K \longrightarrow X^{**}$ the inclusion; let τ denote the topology on X, quotient via ξ of the (weak X^*)-topology on X^{**}; and write τ_ξ for the trace of τ on dX. The following are equivalent for an algebraic linear functional f on X.

(i) f is bounded and $f\circ\xi = i_{X^*}(f)$ $(= f^{**} \in X^{***})$.

(ii) $f \in K^\perp$ $(= \ker(j^t) = \ker(X^* \xrightarrow{i_{X^*}} X^{***} \xrightarrow{j^*} K^*))$.

(iii) f is bounded and $i_{X^*}(f)$ $(= f^{**})$ vanishes on K.

(iv) f is τ-continuous.

(v) f is τ_ξ-continuous on dX.

(vi) $f\circ\xi$ is (weak X^*)-continuous on X^{**}.

(vii) $f\circ\xi$ is (weak X^*)-continuous on the disc dX^{**}.

PROOF. (i) $\Rightarrow$ (ii) : $j^*(i_{X*}(f)) = j^*(f \circ \xi) = f \circ \xi \circ j = f \circ 0 = 0$.
(ii) $\Rightarrow$ (iii) is trivial, as are (iv) $\Rightarrow$ (v) $\Rightarrow$ (vii) and (iv) $\Rightarrow$ (vi) $\Rightarrow$ (vii).
(iii) $\Rightarrow$ (iv) : Since f^{**} is (weak X^*)-continuous on X^{**} and vanishes on $\ker(\xi)$, it factors uniquely through ξ as $f^{**} = \tilde{f} \circ \xi$ with $\tilde{f}$ τ-continuous on X. But $\xi \circ i_X = id_X$ and $f^{**} \circ i_X = f$, and so $f = f^{**} \circ i_X = \tilde{f} \circ \xi \circ i_X = \tilde{f}$ is τ-continuous.
(vii) $\Rightarrow$ (i) : If $f \circ \xi$ is (weak X^*)-continuous on the (weak X^*)-compact disc dX^{**}, the image $f(dX) = f(\xi(dX^{**}))$ is compact in the scalar field, and f is bounded. Now both f^{**} and $f \circ \xi$ are linear functionals on X^{**}, (weak X^*)-continuous at least on the unit disc dX^{**}, and equal on the (weak X^*)-dense subset $i_X(dX) \subset dX^{**}$, as the calculation $f \circ \xi \circ i_X = f = f^{**} \circ i_X$ shows. But then $f \circ \xi$ and f^{**} agree on the entire disc dX^{**}, hence are equal.

REMARK. Similar reasoning shows that a bounded linear transformation $f: A \longrightarrow B$, where (A, α) and (B, β) are $**$-algebras, is a $**$-homomorphism if and only if it is $(\tau_\alpha, \tau_\beta)$-continuous, or equivalently, if and only if f^{**} carries $\ker(\alpha)$ into $\ker(\beta)$.

Recalling from the proof of (2.2) that the canonical preconjugate associated with a $**$-algebra (X, ξ) is precisely the polar $K^\perp$ of the kernel $K = \ker(\xi)$ of the algebra structure ξ, we may gauge the utility of Lemma 4 by the brevity of the proof it affords for the following, now trivial, observation, which we record here largely for completeness' sake.

PROPOSITION 4. The passage $\xi \mapsto \tau_\xi : Alg(X) \longrightarrow Comp(X)$ is bijective.

PROOF. By the equivalence of (ii) with (v), the relation $\tau_\xi = \tau_{\xi'}$ makes the canonical preconjugates for (X, ξ) and (X, ξ') coincide, whence $\xi = \xi'$.

6. EPILOGUE. In the fallout of Propositions 1, 2, 3, and 4, we find the promised characterization theorems of Dixmier, Waelbroeck, and Ng.

OBSERVATION. Conditions, each necessary and sufficient for a Banach space X to be isometrically isomorphic to a conjugate space, are:

(i) [2: Th. 18] there is a Dixmier complement for X in X^{**};

(ii) [2: Th. 19] there is a norm-closed, (weak X)-dense subspace V of X^* with dX (weak V)-compact;

(iii) [18], [23: Prop. 1] for some separated, locally convex topology τ on X, the unit disc dX is τ-compact;

(iv) [24: §3] (almost) there is a compact Hausdorff topology $\varkappa$ on dX, weaker than the (weak X^*)-topology, in which the subconvex combination operations $(x, y) \mapsto \frac{1}{2}(x+y)$ and $x \mapsto \alpha x$ $(|\alpha| \leq 1)$ are continuous.

PROOF. If any one of the sets $Pre(X)$, $Comp(X)$, $Dix(X)$ (of Dixmier complements for X), or $Pre_c(X)$, each in bijective correspondence with $Alg(X)$, is non empty, so are they all.

BIBLIOGRAPHY

1. S. Banach, Théorie des Opérations Linéaires, Monog. Mat., Warsaw, 1932.

2. J. Dixmier, "Sur un théorème de Banach", Duke Math. J., 15 (1948), 1057-1071.

3. N. Dunford and J. T. Schwartz, Linear Operators, Part I: General Theory (P. A. M. 7), Interscience, N. Y., 1958.

4. S. Eilenberg and G. M. Kelly, "Closed categories", Proc. Conf. Categ. Algebra (La Jolla, 1965), Springer, Berlin, 1966, pp. 421-562.

5. S. Eilenberg and J. C. Moore, "Adjoint functors and triples", Illinois J. Math., 9 (1965), 381-398.

6. G. J. O. Jameson, Topology and Normed Spaces, Chapman/Hall, London, 1974.

7. J. L. Kelley, I. Namioka, et al., Linear Topological Spaces, Van Nostrand, Princeton, 1963.

8. F. E. J. Linton, "Some aspects of equational categories", Proc. Conf. Categ. Algebra (La Jolla, 1965), Springer, Berlin, 1966, pp. 84-94.

9. F. E. J. Linton, "An outline of functorial semantics", Springer L. N. M., 80 (1969), 7-52.

10. F. E. J. Linton, "Applied functorial semantics, II", Springer L. N. M., 80 (1969), 53-74.

11. F. E. J. Linton, "Relative functorial semantics: adjointness results", Springer L. N. M., 99 (1969), 384-418.

12. F. E. J. Linton, "Preliminary report on a pullback lemma for Banach spaces and the functorial semantics of double dualization", preprint, Dalhousie University, Halifax, Nova Scotia, 1970.

13. F. E. J. Linton, "Addendum to '... functorial semantics of double dualization'", preprint, Dalhousie University, Halifax, Nova Scotia, 1970.

14. F. E. J. Linton, "Three categorical novelties", Notices Amer. Math. Soc., 19 (1972), A-387.

15. W. A. J. Luxemburg, "On closed linear subspaces and dense linear subspaces of locally convex topological linear spaces", Proc. Int. Symp. Linear Spaces (Jerusalem, 1960), Jerusalem Acad. Press, Jerusalem, 1961, pp. 307-318.

16. S. Mac Lane, Categories for the Working Mathematician, Springer, 1971.

17. E. G. Manes, Algebraic Theories, Springer, N. Y., 1976.

18. Kung-Fu Ng, "On a theorem of Dixmier", Math. Scand., 29 (1971), 279-280.

19. H. H. Schaeffer, Topological Vector Spaces, Macmillan, N. Y., 1966.

20. Z. Semadeni, Monads and their Eilenberg-Moore Algebras in Functional Analysis, Queen's Papers P. A. M., 33, Queen's Univ., Kingston, Ontario, 1973.

21. T. Świrszcz, Monadic Functors and Convexity, dissertation, Warsaw, 1974.

22. T. Świrszcz, Monadic Functors and Categories of Convex Sets, Preprint No. 72, Inst. Math. Polish Acad. Sci., Warsaw, 1975.

23. L. Waelbroeck, "Duality and the injective tensor product", Mathem. Annalen, 163 (1966), 122-126.

24. L. Waelbroeck, "Some theorems about bounded structures", J. Functional Analysis, 1 (1967), 392-408.

DEPARTMENT OF MATHEMATICS
WESLEYAN UNIVERSITY
MIDDLETOWN, CONN. 06457

Contemporary Mathematics
Volume 2
1980

STONIAN DIFFERENTIATION AND REPRESENTATION OF VECTOR FUNCTIONS AND MEASURES

by

Dennis Sentilles

In this paper we show that countably additive finite measures ν, μ on σ-complete Boolean algebras may be differentiated one by the other at <u>every</u> point of the associated Stone representative space according to the formula

$$\lim_{a \searrow s} \frac{\nu(a)}{\mu(a)} = D_\mu \nu(s)$$

where $a \searrow s$ through the neighborhood filter of clopen sets a containing s, and s is in the support of μ.

In Section 2 we examine the function defined at <u>each</u> $s \in S$ by

$$DT(s) = w^* - \lim_{a \searrow s} \frac{T(\mu_a)}{\mu(a)} \in X^{**}$$

where $T: L^1(\mu) \to X$ is a bounded operator into the Banach space X. When X has RNP, DT is $\| \ \|$-continuous, X-valued and the $\| \ \|$-limit of $T(\mu_a)/\mu(a)$, all these except on a nowhere dense subset of S, and this characterizes RNP. We achieve this by a new and simple proof of the Dunford-Pettis-Phillips theorem on weakly compact operators on L^1, **and a sharp distinction, via S, between those functions DT arising from a weakly compact operator and those arising from a Bochner representable operator. The key here is a representation of the spaces $L^\infty(\Omega, X, \mu)$ as a space of continuous functions on S.**

In Section 3 we address the question: What does the number $D_\mu \nu(s)$ represent or mean? In so doing we find a technique which yields a trivial proof of the lifting theorem for Lebesgue measure on the line.

<u>Preliminaries</u>. Henceforth $\mathcal{A}$ denotes a σ-complete Boolean algebra, and S denotes

Copyright © 1980, American Mathematical Society

its Stone representation space. All other needed notation and definitions can be found in the introduction to [12] with the exception that herein $a \in A$ will denote both the algebra element a and its clopen representative (which we again call a) in S. Another convention is, for example, that we shall write $a \subset b$ in a topological argument and $a \leq b$ in an algebraic one. The space S is of course compact and totally disconnected and the sets a, $a \in A$ and $s \in a$, form a clopen neighborhood basis at the point $s \in S$.

A measure μ will be called strictly positive, written $\mu >> 0$, if $\mu(a) > 0$ for all $a \in A$, $a \neq 0$. In such a case, A and S meet the countable chain condition (c.c.c.): that any pairwise disjoint (p.w.d.) subset B of A is at most countable,(because $\sup\{ \sum_{b \in B'} \mu(b) : B' \text{ finite}\} \leq \mu(e) < \infty$, e being the identity in A).

When there exists $\mu >> 0$ on A, then A being already σ-complete, is by c.c.c. complete: any $B \subset A$ has a supremum in A [7]. Equivalently, S is extremely disconnected: i.e., the closure of any open set is open. Moreover, again under all these conditions, each set of first category in S is nowhere dense [8, Theorem 22.2].

The ready example of these phenomena is given as follows. Let $(\Omega, \Sigma, \tilde{\mu})$ be a finite measure space, Σ a σ-field of subsets of Ω. Let $A = \Sigma/\tilde{\mu}^{-1}(0)$ be the quotient Boolean algebra mod the σ-ideal of sets of $\tilde{\mu}$ measure zero. Define μ on A by $\mu(a) = \tilde{\mu}(a)$ where $a = A + \tilde{\mu}^{-1}(0)$. Then $\mu >> 0$. Notice that if $\mu(\{\omega\}) = 0$ for $\omega \in \Omega$, the points of Ω are lost in A and in the associated S. To save much writing, whenever this setting is under discussion we will always let $B, A \in \Sigma$ correspond to $b, a \in A$, replace $\tilde{\mu}$ by μ, and also write $\mu(A) = \mu(a)$. This example is further discussed in [13] and we want to use the notation therein and embodied in [13, Thm. 4.7]. In particular, $L^1(A)$ denotes $L^1(\Omega,\Sigma,\mu)$, and μ_a denotes $\chi_A \in L^1(\Omega,\Sigma,\mu)$, where $\mu_a(b) = \mu(ab)$.

In this same setting we emphasize too that $L^\infty(\Omega,\Sigma,\mu)$ may be identified with the bounded continuous real-valued functions on S: if the equivalence class $[f] \in L^\infty(\Omega,\Sigma,\mu)$, we will write $\hat{f}$, or even f when the meaning is clear, to denote the associated function in $C(S) \equiv L^\infty(A)$ [13] and write $L^\infty(A)$ for $L^\infty(\Omega,\Sigma,\mu)$. The map $[f] \to \hat{f}$ is not easily defined by a point mapping of Ω into S (see [14,

p. 207]) because points of Ω are lost in S. It is this which we address in Section 3. A consequence of Section 1 is that $\hat{f}(s) = \lim_{a \searrow s} \frac{\int_A f d\mu}{\mu(A)}$ at each $s \in S$. The example of Lebesgue measure on $\Omega = [0,1]$ will suffice for the entire discussion.

Finally, μ on $\mathcal{A}$ has a Borel representation $\bar{\mu}$ on S [13, Thm.4.6]. If $[f] \in L^\infty(\Omega,\Sigma,\mu)$, $\int_a \hat{f}\, d\bar{\mu} = \int_A f d\mu$ and we shall replace $\bar{\mu}$ by μ and write $\int_a \hat{f}\, d\mu$ for integrals over subsets a of S and thus also write $\int_a \hat{f}\, d\mu = \int_A f\, d\mu$. As long as expressions involve integration or measure, A and a and [f] and $\hat{f}$ are indistinguishable. If ν is a measure on $\mathcal{A}$ and $f \in C(S)$ we shall also write $\nu(f) = \int_S f\, d\nu$, especially when we wish to regard ν as a linear functional on C(S). It is shown in [13] that $L^1(\mathcal{A})$ identifies with the finite measures on $\mathcal{A}$ and is the dual of $L^\infty(\mathcal{A})$ in the strict topology.

1. Differentiation of measures. The principal result of this section, Theorem 1.2, **appears** in [6] where it was derived in order to describe the dual of the space of measures. Here it is a central result and we will include its simple proof so as to make this paper self-contained and to make the result available to one who is not interested in the details of [6]. The only background needed for this section is point set topology, induction on $[1,\omega_1)$, and a willingness to exploit the duality of Boolean elements a and their Stone representatives. All measures are finite valued and countably additive, and (again) $\mathcal{A}$ is σ-complete. The following lemma is the Boolean analogue of the Hahn-Jordan decomposition of measures on σ-fields. A proof can be found in [3].

Lemma 1.1. If γ is a measure on $\mathcal{A}$ and $\gamma(a)>0$ for some $a \in \mathcal{A}$, then there exists $b \in \mathcal{A}$ such that $\gamma_b \geq 0$ on $\mathcal{A}$. Moreover we can find b so that $b \leq a$, and especially, $b \neq 0$.

Here of course $\gamma_b(c)$ means $\gamma(bc)$ and $\gamma_b \geq 0$ means $\gamma_b(c) \geq 0$ for all $c \in \mathcal{A}$.

Theorem 1.2. Let $\mu >> 0$ on $\mathcal{A}$ and ν be a non-negative measure on $\mathcal{A}$.

If $s \in S$, then

$$D_\mu \nu(s) = \lim_{a \searrow s} \frac{\nu(a)}{\mu(a)}$$

exists, where the limit is taken over the neighborhood filter of clopen sets containing s.

Proof: Suppose $\underline{\lim}_{a \searrow s} \frac{\nu(a)}{\mu(a)} < \sigma < \overline{\lim}_{a \searrow s} \frac{\nu(a)}{\mu(a)}$. We will produce a p.w.d. subset of distinct elements $a_\alpha \in A$ indexed by $\alpha \in [1,\omega_1)$, where ω_1 is the first uncountable ordinal, which will contradict c.c.c.

Find $b_1 \ni s$ such that $\sigma < \frac{\nu(b_1)}{\mu(b_1)}$. Then $(\nu - \sigma\mu)(b_1) > 0$. Hence by 1.1, there exists $a_1 \neq 0$, $a_1 \leq b_1$ such that $(\nu - \sigma\mu)_{a_1} \geq 0$. But this means $s \notin a_1$. Because if $s \in a_1$, then for all $a \subset a_1$, $s \in a$, we have $(\nu - \sigma\mu)(a) = (\nu - \sigma\mu)(aa_1) = (\nu - \sigma\mu)_{a_1}(a) \geq 0$ so that $\underline{\lim}_{a \searrow s} \frac{\nu(a)}{\mu(a)} \geq \sigma$.

Hence $s \notin a_1$. Since a_1 is closed pick $b_2 \ni s$ such that $\sigma < \frac{\nu(b_2)}{\mu(b_2)}$ and $b_2 a_1 = 0$. Find $a_2 \neq 0$, $a_2 \leq b_2$ so that $(\nu - \sigma\mu)_{a_2} \geq 0$. Then $a_2 a_1 = 0$ and $s \notin a_2$.

Suppose we have an ordinal $\xi < \omega_1$ and that for all $\alpha < \xi$ we have $a_\alpha \neq 0$ such that (1) $s \notin a_\alpha$ (2) $(\nu - \sigma\mu)a_\alpha \geq 0$ (3) $a_\alpha a_\gamma = 0$ for $\alpha \neq \gamma$. We find a_ξ as follows. Let $C_\xi = \bigvee_{\alpha<\xi} a_\alpha$; topologically, $C_\xi = \overline{\bigcup_{\alpha < \xi} a_\alpha}$. We claim $s \notin C_\xi$. For, C_ξ is clopen and if $s \in C_\xi$, then $s \in a \subset C_\xi$ implies

$$(\nu - \sigma\mu)(a) = (\nu - \sigma\mu)(a\, C_\xi) = (\nu - \sigma\mu)_a(C_\xi)$$

$$= (\nu - \sigma\mu)_a\Big(\bigvee_{\alpha<\xi} a_\alpha\Big) = \sum_{\alpha<\xi} (\nu - \sigma\mu)_a(a_\alpha) \geq 0,$$

the last equality because ξ is a countable ordinal and $(\nu - \sigma\mu)_a$ is countably additive and the $\{a_\alpha\}_{\alpha<\xi}$ are p.w.d. But then, $\underline{\lim}_{a \searrow s} \frac{\nu(a)}{\mu(a)} \geq \sigma$.

Hence $s \notin C_\xi$ and since C_ξ is also closed we find $b_\xi \cap C_\xi = \square$ and $(\nu - \sigma\mu)(b_\xi) \geq 0$ and from this find a_ξ such that $(\nu - \sigma\mu)_{a_\xi} \geq 0$ with $a_\xi a_\alpha = 0$ for all $\alpha < \xi$. This completes the inductive argument and yields a contradiction and the proof.

Theorem 1.3. $D_\mu \nu$ is continuous on S and $D_\mu \nu$ is finite valued except on a closed nowhere dense set in S.

Proof: Suppose $\nu \geq 0$. Let $a_n = \sup\{b \in A: (n\mu - \nu)_b \geq 0\}$. Clearly a_n increases in A and if $s \in a \subset a_n$ then $\frac{\nu(a)}{\mu(a)} \leq n$ so that $D_\mu \nu \leq n$ on a_n. It is clear that $a_n \nearrow e$ since ν is finite valued. Hence $D_\mu \nu$ is finite on $\bigcup_{n=1}^{\infty} a_n$ which is dense and open in S.

For continuity, if $D_\mu \nu(S) = +\infty$, then $D_\mu \nu > n$ on the clopen set $S\backslash a_n$. Suppose $D_\mu \nu(S) = \alpha$ and $\epsilon > 0$. Find b, c such that $((\alpha+\epsilon)\mu - \nu)_b \geq 0$ and $((\alpha-\epsilon)\mu - \nu)_c \geq 0$. Since $D_\mu \nu(s) = \alpha$ it follows that $s\epsilon b\backslash c$ and if $t\epsilon a \subset b\backslash c$, then

$$\alpha - \epsilon \leq \frac{\nu(a)}{\mu(a)} \leq \alpha+\epsilon \ .$$

Consequently, $\alpha-\epsilon \leq D_\mu \nu(t) \leq \alpha+\epsilon$ for all $t\epsilon\ b\backslash c$, and $D_\mu \nu$ is continuous at s.

In the case where ν is not non-negative, 1.1 allows one to write $\nu = \nu^+ - \nu^-$, $\nu^+, \nu^- \geq 0$. Then we can define $D_\mu \nu$ at all points of s because we can also obtain $\nu^+ = \nu_a$, $\nu^- = \nu_b$ for some a and b with $ab = 0$ and both clopen. This completes the proof.

Corollary 1.4. If μ, ν are measures on A and $\mu \geq 0$ then

$$\lim_{a \searrow s} \frac{\nu(a)}{\mu(a)}$$

exists and is continuous in the relative topology at each point s in the support $S_\mu \subset S$ of the corresponding Borel measure μ on S.

Proof. Again suppose $\nu \geq 0$. It is a matter of manipulating definitions to show that if $B = A/\mu^{-1}(0)$ then S_μ is the Stone space of B and the Stone topology is the relative topology on S_μ. The Boolean analogue of the Lebesgue Decomposition Theorem (see [3], [5], or [6]) allows one to write $\nu = \nu_0 + \nu_1$ where ν_0 vanishes on $\mu^{-1}(0)$ and $S_\nu \cap S_{\nu_1} = \square$ (here ν_1 is the corresponding Borel measure on S). If $\tilde{\mu}$ and $\tilde{\nu}_0$ represents the corresponding measures on B, then for the neighborhood $a \cap S_\mu$ of $s\epsilon S_\mu$ one has

$$\tilde{\nu}_0(a\cap S_\mu) = \nu_0(a) \text{ and } \tilde{\mu}(a\cap S_\mu) = \mu(a).$$

By, 1.2 $\lim_{a \searrow s} \nu_0(a)/\mu(a)$ exists. Now $\lim_{a \searrow s} \frac{\nu_1(a)}{\mu\ (a)} = 0$ because the support of ν_1 in S is closed and disjoint from S_μ, and $s\epsilon S_\mu$ then implies that there is a neighborhood a of s such that $\nu_1(a) = 0$. Hence $\lim_{a \searrow s} \frac{\nu(a)}{\mu(a)}$ exists and being equal to $\lim_{a \searrow s} \frac{\tilde{\nu}(a\cap S_\mu)}{\tilde{\mu}(a\cap S_\mu)}$ is continuous on S_μ by 1.3.

Remark: We would like to point out here the especially simple proof of the Lebesgue Decomposition obtained by Graves [5]. Simply let

$$\nu_1(b) = \sup \{\nu(c) : c \in b\cdot\mu^{-1}(0)\}$$

Because $\mu^{-1}(0)$ is an ideal in A, it readily follows that ν is countably additive, and if $\mu(b) = 0$, then $\nu_1(b) = \nu(b)$ so $\nu - \nu_1 \ll \mu$.

Employing the identification of $L^\infty(\Omega,\Sigma,\mu)$ with $C(S)$ and the usual σ-finite argument, along with entirely formal manipulations of integrals $\int_A f\, d\mu$ and $\int_{\hat{a}} f\, d\mu$, the results above include the standard Radon-Nikodym theorem and the result that $L^{1*} = L^\infty$

Corollary 1.5. Let $\mu \gg 0$ on A. (a) For all $a \in A$, $\nu(a) = \int_a D_\mu \nu d\mu$ as an integral in S. Hence $\nu(A) = \int_A f\, d\mu$ where $[f] \in L^\infty(\Omega)$ and $\hat{f} = D_\mu \nu$

(b) If $\phi \in L^1(A)^*$, then

$$\phi(\nu) = \int_S \hat{\phi}\, d\nu$$

where $$\hat{\phi}(s) = \lim_{a \searrow s} \frac{\phi(\mu_a)}{\mu(a)} \in C(S)$$

Consequently, $L^1(\Omega)^* = L^\infty(\Omega)$.

The details of 1.4 can be found in [6], but it is also a good exercise. Note that $|\phi(s)| < \infty$ for all s because

$$\left| \frac{\phi(\mu_a)}{\mu(a)} \right| = \left| \phi\left(\frac{\mu_a}{\mu(a)}\right) \right| \le \|\phi\| \left\| \frac{\mu_a}{\mu(a)} \right\|_1 \le \|\phi\|$$

As a second application of 1.2 and 1.3 we give an easy proof of a well-known result.

Corollary 1.6. Let $\mu \gg 0$ on A and $\mathcal{D}$ be a directed set of non-negative functions in $C(S)$ which is bounded above by some $h \in C(S)$. Then the lattice supremum $\bigvee \mathcal{D}$ exists in $C(S)$.

Proof. Define ν on A by

$$\nu(a) = \sup \{\textstyle\int_a f\, d\mu : f \in \mathcal{D}\}$$

Then ν is bounded by $\int_S h\, d\mu$ and ν is finitely additive. But if a_n a in A, then

$$\sup_n \nu(a_n) = \sup_n \sup_{f\in\mathcal{D}} \int_{a_n} f\, d\mu$$

$$\ge \sup_{f\in\mathcal{D}} \sup_n \int_{a_n} f\, d\mu$$

$$\ge \sup_{f\in\mathcal{D}} \int_a f\, d\mu = \nu(a)$$

Hence ν is countably additive. It follows that $D_\mu \nu$ is the (lattice) supremum of $\mathcal{D}$ in $C(S)$.

We should note however that this result is known more generally, for any externally disconnected compact S [3 , 43D].

If f is a continuously differentiable function on [0, 1] and $\nu(E) = \int_E f'\,dm$, m being Lebesgue measure, then with $\Delta = (x - \delta, x + \delta)$ the limit $f'(x) = \lim\limits_{\Delta \downarrow x} \frac{\nu(\Delta)}{\mu(\Delta)}$ has of course a well-known meaning. In Section 3 we will try to find a meaning for $\lim\limits_{a \downarrow s} \frac{\nu(a)}{m(a)}$ for the Boolean algebra of Lebesgue sets mod sets of measure zero.

2. Differentiation of vector measures and the Radon-Nikodym property for Banach spaces.

In this section we investigate the limit

$$\lim_{a \downarrow s} \frac{\nu(a)}{\mu(a)}$$

when ν takes its values in a Banach space X and is countably additive. The basic result is the existence of a weak limit, from Section 1, and a variant of the Dunford-Pettis-Phillips theorem which we shall prove: If $f: S \to X$ is continuous with the weak topology, then f(S) is separable and, except on a closed nowhere dense set, f is continuous with the norm topology on X.

At the start I would like to thank B. J. Pettis and W. H. Graves for sparking my interest in this subject, and credit [1] for making a painless introduction possible. I would also like to thank Ren Tai Kuo for many helpful discussions on the topics of this section.

Because we wish to investigate the Radon-Nikodym property in the Stone space context and hence deal with continuous real-valued and necessarily bounded functions, we avoid technical details at the outset by appeal to the basic but simple result [1 , III, 1.5] that X has RNP iff every bounded operator $T: L^1(\Omega,\Sigma,\mu) \to X$ is Bochner representable in $L^\infty(\Omega,X,\Sigma,\mu) \equiv L^\infty(X,\mu)$. That is there is a unique $[g] \in L^\infty(X,\mu)$ such that $Tf = \text{Bochner} - \int_\Omega fg\,d\mu$ for all $f \in L^1(\mu)$ and $\|T\| = \|g\|_\infty$. As a result we can confine our study to bounded operators on $L^1(\mu)$ and yet be assured that RNP is implicitly discussed.

If E and F are locally convex linear spaces in duality, $\sigma(E,F)$ denotes the weak topology on E defined by (pointwise convergence on) F; our reference for these matters is [10], but we will write X^* rather than X' in deference to [1]. The notations of previous sections, in particular μ_a and $L^1(A)$ are maintained here. We will also have need for the notation $\chi(a)$, $a \in A$ [13], which denotes the continuous 0-1 valued function $\chi_a \in C(S)$.

Let then $T: L^1(\mu) \to X$ as above. For each $x^* \in X^*$, $\nu(a) = x^* T(\mu_a)$ is a countably additive measure on A and the limit

$$f_{x^*}(s) = \lim_{a \searrow s} \frac{x^* T\mu_a}{\mu(a)}$$

exists at each $s \in S$ and is continuous by Theorem 1.3. Set $f(s)(x^*) = x^* f(s)$. Then $| x^* T(\mu_a) | \le || x^* || \, || T || \, || \mu_a ||_1 = || x^* || \, || T || \mu(a)$ so that $f: S \to X^{**}$ and f is $\sigma(X^{**}, X^*)$ continuous. Henceforth we denote the function f so defined by DT. That is

$$DT(s) = \text{weak}^*\text{-} \lim_{a \searrow s} \frac{T(\mu_a)}{\mu(a)} \text{ in } X^{**}.$$

Denote by $C_o(S, X_\sigma)$ the set of functions $g: S \to X^{**}$ such that $g(s) \in X$ except for s in a closed nowhere dense subset of S and such that g is continuous in the $\sigma(X^{**}, X^*)$ topology on X^{**}. Let $C(S, X_\sigma)$ denote the set of those g in $C_o(S, X_\sigma)$ such that $g(S) \subset X$ and let $C(S, X)$ denote those norm continuous g in $C(S, X_\sigma)$. Let $|| g || = \sup \{|| g(s) ||: s \in S\}$; each of these spaces is a $|| \; ||$-closed subspace of that which precedes it above. We will see that T is compact (weakly compact) (Riesz representable [1]) iff DT lies in $C(S, X)(C(S, X_\sigma))(C_o(S, X_\sigma))$, and that any such DT is $|| \; ||$-continuous except on a closed nowhere dense set in S. Recall again that in the Stone space S (of a measure algebra) each set of first category is again nowhere dense.

Now is perhaps a good point at which to address an important point often reiterated in Diestel and Uhl [1]: that a representation theory for operators on L^1 into X must have content and it is the Bochner integral that carries the content. It is a trivial matter, using Section 1 and 1.5 to obtain a weak integral representation of T by way of DT(s). Most of the work of this section is devoted to showing that DT carries a Bochner representation. More precisely we will show that the analysis of T and of DT can be carried on

topologically in S and in $C_o(S, X_\sigma)$ but that carried along as well is an underlying Bochner theory on $L^\infty(\Omega,X,\Sigma,\mu)$. At the same time we will avoid technical difficulties by only dealing with weak integrals and Borel measures on S.
We will do this by identifying $L^\infty(X, \mu)$ with $C_o(S, X_\sigma)$ by a surjective isometry $[f] \to \hat{f}$, such that for $x^* \in X^*$ one has

$$x^*(\text{Bochner} - \int_\Omega f\, d\mu) = \int_S x^* \hat{f}\, d\mu.$$

First of all, the space C(S, X) is far too small for our needs.

Lemma 2.1. If $g \in C(S, X)$ and $\epsilon>0$, then there exists $x_1 \ldots, x_n \in X$ and $a_1, \ldots, a_n \in A$ such that $\| g(s) - \sum_{i=1}^{n} x_i\, \chi(a_i)(s) \| < \epsilon$ for all $s \in S$.

Proof: Since g(S) is $\| \ \|$-compact there exists $x_1, \ldots, x_n$, $x_i = g(s_i)$ such that $\inf_k \| g(s) - x_k \| < \epsilon$ for each $s \in S$. The set $U_i = \{s: \|g(s) - x_i\| < \epsilon\}$ is open in S. Let $b_i \in A$ denote its closure. Let $a_k = b_k - \sum_{i=1}^{k-1} b_i$. Pick $t_k \in a_k$ and note that $\|g(s) - \sum_{k=1}^{n} g(t_k)\, \chi(a_k)(s) \| \le 2\epsilon$ for all s, to complete the proof.

To go on we begin the task of assuring the reader that this setting includes the standard theory of $L^\infty(\Omega,X,\Sigma,\mu)$. Let $S(\Sigma,X)$ represent the simple X-valued Σ-measurable functions on Ω, and $S(A,X)$ the subspace $\{ \sum_{i=1}^{n} x_i\, \chi(a_i): x_i \in X,\ a_i \in A\}$ of C(S, X). As in [13 , Thm. 4.7] the function $\chi_A \in L^\infty(\Omega,\Sigma,\mu)$ is identified with $\chi(a) \in S(A)$, where $a = A + \mu^{-1}(0)$ in $\Sigma/\mu^{-1}(0)$. For this part of the discussion we write $\hat{\chi}_A = \chi(a)$ and for $[f] \in L^\infty(\Omega,\Sigma,\mu)$ let $\hat{f}$ denote the corresponding function in $C(S) \equiv L^\infty(A)$; recall that $\| f \|_\infty = \| \hat{f} \|$ in C(S) [13 , p. 11].
Now for $[f] \in L^\infty(X, \mu)$ set

$$\hat{f}(s)(x^*) = \widehat{x^* f}(s).$$

We will now begin to simply write f for the equivalence class [f]. Because f is bounded, $\hat{f} \in C(S,(X^{**}, \sigma(X^{**}, X^*))$. In fact

Theorem 2.2. (a) If $f = \sum_{i=1}^{n} x_i \chi_{A_i}$, then $\hat{f} = \sum_{i=1}^{n} x_i \chi(a_i)$

(b) If $f \in L^\infty(\Omega,X,\Sigma,\mu)$, then $\hat{f} \in C_o(S, X_\sigma)$

(c) The mapping $f \to \hat{f}$ is an isometry.

Proof: (a) $\hat{f}(s)x^* = \widehat{x^*f}(s) = (\widehat{\sum_{i=1}^{n} x^*(x_i)\chi_{A_i}})(s)$

$$= \sum_{i=1}^{n} x^*(x_i)\ \chi(a_i)(s) = x^*\ (\sum_{i=1}^{n} x_i\ \chi(a_i)(s))$$

because $\widehat{\sum_{i=1}^{n} \alpha_i\chi_{A_i}} = \sum_{i=1}^{n} \alpha_i\chi(a_i)$ in the identification of $L^\infty(\Omega,\Sigma,\mu)$ with $C(S)$.

(c) Trivially, if $f \in S(\Sigma, X)$, $||\ f\ || = ||\ \hat{f}\ ||$

Given $f \in L^\infty(X, \mu)$ and any $||x^*|| \le 1$, one has

$$||\ f\ ||_\infty \ge ||x^*f\ ||_\infty = ||\widehat{x^*f}\ || = \sup\ \{|x^*f(s)|: s\in S\}.$$

Given $\epsilon>0$ find $s\in S$ such that $||\ \hat{f}\ || - \epsilon < ||\ \hat{f}(s)||$

and $||x^*|| \le 1$ such that $||\ \hat{f}(s)|| - \epsilon < |\ x^*\hat{f}(s)| = |\ \widehat{x^*f}(s)| \le ||\ f\ ||_\infty$

Hence $||\ f\ ||_\infty > ||\ \hat{f}\ ||$. On the other hand, if $\omega\in\Omega$,

$$\begin{aligned} ||f(\omega)|| &= \sup\ \{|(x^*f)(\omega)|: ||x^*|| \le 1\} \\ &\le \sup\ \{||x^*f||_\infty: ||x^*|| \le 1\} \\ &= \sup\ \{||\widehat{x^*f}||: ||x^*|| \le 1\} \\ &= \sup\ \{\sup_S\ |\widehat{x^*f(s)}|: ||x^*|| \le 1\} \\ &\le \sup\ ||\hat{f}(s)|| \\ &= ||\ \hat{f}\ || \end{aligned}$$

So $||\ f\ ||_\infty = ||\ \hat{f}\ ||$.

(b) If $f\in L^\infty(X, \mu)$ find $g = \sum_{n=1}^{\infty} x_n\chi_{E_n}$, E_n p.w.d., such that $||f - g||_\infty < \epsilon$

Since $C_o(S, X_\sigma)$ is $||\ ||$-closed, it suffices to obtain $\hat{g}\in C_o(S, X_\sigma)$. Write $\hat{\chi}_{E_k} = \chi(e_k)$. We may assume $\bigcup_1^\infty E_k = \Omega$ or $\bigvee_1^\infty e_k = e$; hence $S\setminus\bigcup_{k=1}^\infty e_k$ is closed and nowhere dense. Now the e_k are p.w.d. and since $||(g - x_k)\chi_{E_k}\ ||_\infty = 0$ so does $||(\hat{g} - x_k)\chi(e_k)|| = 0$, whence $\hat{g} = x_k$ on e_k. Hence $\hat{g}(s)\in X$ except on $S\setminus\bigcup_{k=1}^\infty e_k$, completing the proof.

It may be helpful to point out the following:

If $\{a_k\}\subset A$ is p.w.d. and $\bigvee a_k = e$, and if $\{x_k\}\subset X$ and bounded, the function $g: \bigcup_{k=1}^\infty a_k \to X \subset X^{**}$ defined by $g(s) = \sum_{k=1}^{\infty} x_k\ \chi(a_k)(s)$ is continuous into a bounded, hence $\sigma(X^{**}, X^*)$ compact subset of X^{**}. Since the open dense subset $\bigcup_1^\infty a_n$ is K-embedded because S is extremally disconnected, g has a unique continuous extension to S into the $\sigma(X^{**}, X^*)$-compact set $\overline{\{x_k\}}^{\sigma(X^{**}, X^*)}$.

I wish to thank Ren Tai Kuo for first pointing out to me that such functions $\hat{g}$ lie in $C_o(S, X_\sigma)$. The next observation insures that Bochner representation on Ω results from the ensuing discussion.

Lemma 2.3. If $f \in L^\infty(\Omega, X, \Sigma, \mu)$ then for any $x^* \in X^*$

$$x^*(\int_\Omega f\, d\mu) = \int_S x^* \hat{f}\, d\mu.$$

where $\int_\Omega f d\mu$ is the Bochner integral of f, and the second integral is the Lebesgue integral of $x^* \hat{f}$ by the Borel representative of μ on S.

Proof: It is well known that $x^* \int_\Omega f\, d\mu = \int_\Omega x^* f\, d\mu$, the latter a Lebesgue integral. But for $g \in L^\infty(\Omega, \Sigma, \mu)$, $\int_\Omega g\, d\mu = \int_S \hat{g}\, d\mu$ [12] and $\widehat{x^* f} = x^* \hat{f}$ by definition of $\hat{f}$.

We will later prove that the map $f \to \hat{f}$ is onto. This depends on the principal result of this section which we now begin to develop.

Lemma 2.4. If $f \in C_o(S, X_\sigma)$, then $s \to ||f(s)||$ is continuous except on a nowhere dense set.

Proof. Since $||f(s)|| = \sup\{|x^* f(s)| : ||x^*|| \le 1\}$, $s \to ||f(s)||$ is lower semi-continuous. But if g is l.s.c., then g is the pointwise supremum of $\{h \in C(S) : h \le g\}$. This latter family, being directed upward, has, by 1.4, a supremum $\xi \in C(S)$ and $\xi \ge g$. And now, $\xi = g$ except on a first category, hence nowhere dense set, by a standard argument: for each n, $\{s \in S : \xi(s) > g(s) + \frac{1}{n}\}$ is nowhere dense.

Theorem 2.5. If $f \in C_o(S, X_\sigma)$ and $f(S \setminus N)$ is $||\ ||$-separable in X, where N nowhere dense in S, then f is $||\ ||$-continuous except on a (possibly different) nowhere dense set.

Proof. By 2.4, $||f - x_k||$ is continuous except on N_k where we suppose $\{x_k\}$ is $||\ ||$ - dense in $f(S \setminus N)$. Let $s \in S \setminus (\bigcup_{k=1}^\infty N_k \cup N)$ and $\epsilon > 0$. Find x_n such that $||f(s) - x_n|| < \epsilon/2$. Then $||f - x_n||$ is continuous at s and so there is a neighborhood a of s such that $t \in a$ implies $||f(t) - x_n|| < \epsilon/2$, whence $||f(t) - f(s)|| < \epsilon$.

Notice that by exactly the same methods as 2.4 and 2.5,

Theorem 2.6. If $f: S \to X$ is weak* -continuous, where X is a separable dual space, then f is $||\ ||$-continuous except on a nowhere dense set.

To achieve the main result we first record the trivial and probably well-known observation.

Lemma 2.7. If $K \subset X$ is weakly compact and there exists a sequence $\{x_n^*\} \subset X^*$ which separates points of K, then K is $\|\ \|$-separable.

Proof. We can suppose $\|x_n^*\| \leq 1$ and define the metric $d: K \times K \to R$ by $d(x, y) = \sum_{n=1}^{\infty} \frac{1}{2^n} |x_n^*(x - y)|$. The identity map on K into (K, d) is clearly weakly continuous, hence K, being weakly compact, is weakly metrizable, hence weakly separable and so norm separable by the Hahn Banach theorem.

The Dunford-Pettis-Phillips theorem is, according to [1], a cornerstone of the study of the Radon-Nikodym property for Banach spaces. We will prove what we believe should be regarded as a Boolean/Stone version of this result. The proof relies on but one fact, that (in D. H. Fremlin's terminology [9]) a Banach space in its weak topology is angelic: every point in the weak closure of a relatively weakly compact set is the limit of a sequence in the set (Eberlein-Smulian).

Let $T: L^1(\Omega,\Sigma,\mu) \to X$ be a weakly compact operator, and let $DT(s) = \text{weak}^* - \lim_{a \searrow s} \frac{T(\mu_a)}{\mu(a)}$. Since T is weakly compact $DT \in C(S, X_\sigma)$. Moreover, it is clear that if $x^* \in X^*$ then $T^*x^* = x^*(DT) \in C(S) = L^\infty(A) = L^1(\mu)^*$. From purely general arguements it then follows that because T is weakly compact, $K = T^*(\text{Ball } X^*)$ is $\sigma(L^\infty, L^{\infty *})$ compact [10 , p. 154]. Thus the first paragraph of the proof of the next, and main theorem, may be omitted when the function f is DT for T a weak compact operator. These remarks are in order because it is plain that the only thing new in this section is the orchestration, and not the theme.

Theorem 2.8. (Dunford-Pettis-Phillips). If $f \in C_o(S, X_\sigma)$, then f has a norm separable range in X.

Proof: We first consider $f \in C(S, X_\sigma)$. Let $K = \{x^*f: \|x^*\| \leq 1\}$. We clain that K is weakly compact $(\sigma(C(S), C(S)^*))$ in C(S). We have in fact two arguments, both of which are instructive: (1) By Aliaglu's theorem, K is pointwise compact in C(S) and bounded. By Grothendieck's weak compactness theorem for C(S) [2, p. 269] K is weakly relatively compact in C(S) and being pointwise closed is weakly closed hence weak compact. Alternatively: (2) Replace $g \in L^1(\mu)$ by the measure $\nu(a) = \int_A g\, d\mu$ on A and for $x^* \in X^*$ set $x^*(Tg) = \int_S x^* f\, d\nu$. By the Krein-Smulian theorem, since $f(S)$ is weak compact in X, $Tg \in X$ and moreover the induced operator T is weakly compact. Hence $T^*(\text{Ball } X^*) = K$ is weakly compact in $C(S) = L^\infty(A) = L^1(\mu)^*$.

Define d on $K \times K$ by $d(x^*f, y^*f) = \int_S |x^*f - y^*f| d\mu$. Then d is a metric on K because $\mu >> 0$. Let h denote the identity map of K with the weak topology $(\sigma(C(S), C(S)^*))$ to K with the d-topology. We show h is continuous by showing $h(\bar{A}^w) \subset \overline{h(A)}^d$. Because C(S) is angelic in its weak topology, any $x^*f \in \bar{A}^w$ is the weak limit of a sequence $x_n^*f \in A$. Hence x^*f is the pointwise limit of $\{x_n^*f\}$, and the Dominated Convergence theorem yields $x^*f \in \overline{h(A)}^d$. Hence h is a homeormorphism and so there exists $\{x_n^*f\} \subset K$ and weakly dense in K. Hence by the Hahn-Banach theorem, for any x^*, $||x^*|| \le 1$ and $\epsilon > 0$, there exists numbers $\alpha_1, \cdots, \alpha_n$ such that $||x^*f - \sum_{h=1}^{n} \alpha_k x_k^*f|| < \epsilon$ in C(S). It follows that $\{x_n^*\}$ separates points of f(S) and so by Lemma 2.7 f(S) is separable.

Now if $f \in C_o(S, X_\sigma)$, and $f(S) \in X$ except on the nowhere dense set N, choose $a_n \in A$ such that $\bigcap_{n=1}^{\infty} a_n = N$. Then $\chi(e - a_n)f \in C(S, X_\sigma)$ and so has separable range. But then $f(S\backslash N)$, the range of f in X, is separable.

The next result is now purely technical but completes the identification of $C_o(S, X_\sigma)$ with $L^\infty(X, \mu)$ begun in 2.2.

<u>Theorem</u> <u>2.9</u>. If $f \in C_o(S, X_\sigma)$, then there exists $g \in L^\infty(X, \mu)$ such that $\hat{g} = f$. Hence $C_o(S, X_\sigma)$ and $L^\infty(X, \mu)$ are identical as Banach spaces.

<u>Proof</u>: By 2.8 and 2.5, f is $|| \; ||$-continuous except on a closed nowhere dense set N. Choose $a_n \in A$ and p.w.d. such that $S\backslash N = \bigcup_{k=1}^{\infty} a_k$. Let $f_n = f\,\chi(a_n)$. Then $f_n \in C(S, X)$ and by 2.1 there exists functions $f_n^k \in S(A, X)$, $f_n^k = \sum_{i=1}^{P_k} x_i^k \chi(e_i^k)$ such that $||f_n - f_n^k|| \overset{k}{\to} 0$. Let $g_n^k = \sum_{i=1}^{P_k} x_i^k \chi_{E_i^k}$, $e_i^k = E_i^k + \mu^{-1}(0)$, and all $e_i^k \subset A_n$. Then $\hat{g}_n^k = f_n^k$ and so g_n^k is $|| \; ||_\infty$-Cauchy in $L^\infty(X, \mu)$ by 2.2 (c). Let g_n by the limit in $L^\infty(X, \mu)$; it is trivial that $g_n = g_n \chi_{A_n}$, $a_n = A_n + \mu^{-1}(0)$. Clearly, $x^*\hat{g}_n = x^*f_n$, so $\hat{g}_n = f_n$. Since the A_n can be chosen p.w.d., $g = \sum_{n=1}^{\infty} g_n = \sum_{n=1}^{\infty} g_n \chi_{A_n} \in L^\infty(\Omega, \mu)$ and $\hat{g} = f$, completing the proof.

Again alternatively, 2.9 may be proven by choosing a maximal p.w.d., hence countable, collection of clopen sets $\{a_n\}$, $a_n \cap N = \square$, and an s_n in each a_n, and such that $||f(t) - f(s)|| < \epsilon$ on a_n. Then $g = \sum_{n=1}^{\infty} f(s_n)\, \chi_{A_n} \in L^{\infty}(X, \mu)$ and $||\hat{g} - f|| < \epsilon$ where $\hat{g} = \sum_{n=1}^{\infty} f(s_n)\, \chi(a_n)$ in accord with the remarks following 2.2.

We turn now to operator representation and the Radon Nikodym Property for X. Note how 2.8 dramatically retrives an old fact, that $L^1(\mu)$ does not have RNP, because if $T: L^1 \to L^1$ is the identity then $DT(s) = \delta_s$, the unit point mass at s for all $s \in S$, so that $DT(S) \subset X^{**}\backslash X$.

<u>Theorem 2.10</u>. T is representable by a $g \in L^{\infty}(X, \mu)$ iff $DT \in C_o(S, X_{\sigma})$. In this case $\hat{g} = DT$.

Alternatively, X has RNP iff for each countably additive measure $\nu: \mathcal{A} \to X$ of bounded variation the $\sigma(X^{**}, X^{*})$- $\lim_{a \searrow s} \frac{\nu(a)}{\mu(a)}$ in X^{**} lies in X for all but a closed nowhere dense subset of S.

<u>Proof</u>: Suppose T is representable, so that $Tf = \int f\, g\, d\mu$, $g \in L^{\infty}(X,\mu)$. Then $T(\mu_a) = \int_{\Omega} \chi_A\, g\, d\mu$ and hence by 2.3 $x^{*}T(\mu_a) = \int_{\Omega} \chi_A\, x^{*}g\, d\mu = \int_a \widehat{x^{*}g}\, d\mu$. Thus $x^{*}(DT(s)) = \lim_{a\, s} \frac{x^{*}(T\mu_a)}{\mu(a)} = \widehat{x^{*} g}(s) = x^{*}\hat{g}(s)$ But $g \in L^{\infty}(X, \mu)$ so $DT \in C_o(S, X_{\sigma})$ by 2.2 (b).

Conversely, if $DT \in C_o(S, X_{\sigma})$, let $g \in L^{\infty}$ such that $\hat{g} = DT$. For $A \in \Sigma$, $\int \chi_A\, g\, d\mu$ exists as a Bochner integral and by 2.3, $x^{*}(\int \chi_A g\, d\mu) = \int_a \widehat{x^{*}g}\, d\mu = \int_a x^{*}(DT) d\mu$. But by 1.5 (a), with $\nu(a) = x^{*}T\mu_a$, one has $\int_a x^{*}(DT) d\mu = x^{*}(T\mu_a) = x^{*}(T\chi_A)$. Hence $T\chi_A = \int \chi_A\, g\, d\mu$, and the conclusion follows.

For the remainder of the theorem, we merely say that the definition of countable additivity and bounded variation on $\mathcal{A}$ are the same as on σ-fields Σ and that X has RNP iff all $T: L^1(\mu) \to X$ are representable [1, III, 1.5]. The remaining details are purely technical and we omit them.

Finally, as is well known, compact and weakly compact operators are representable, and we obtain this here with the aid of 2.2, 2.3 and 2.10. Notice that (b) and (c) below point up for the first time the ability to differentiate at <u>every</u> point of S.

<u>Theorem 2.11</u>. (a) T is compact on L^1 iff $DT \in C(S, X)$

(b) T is weakly compact iff $DT \in C(S, X_\sigma)$.

(c) X has RNP iff $DT \in C_o(S, X_\sigma)$ for all $T: L^1 \to X$.

(d) If X is a separable dual space, X has RNP.

Proof: (a) If T is compact, then $\{ \frac{T(\mu_a)}{\mu(a)} : a \in A\}$ is relatively norm compact and so $DT \in C(S, X)$. Conversely, the argument in Theorem 2.10 above shows that $T(f)$, for $f \in L^1$, $\| f \|_1 \leq 1$, lies in the closed convex hull of $DT(S)$, so $DT \in C(S, X)$ implies T is compact.

(b) Similarly, if $DT \in C(S, X_\sigma)$, then $T(f)$ is in the closed convex hull of the weakly compact set $DT(S)$, and so T is weakly compact. The converse is now apparent from the first observation in (a) above since now $DT(S)$ lies in the weak closure of the weakly compact set $\{T(\mu_a/\mu(a)): a \in A\}$.

(c) This again is Diestel and Uhl [1, III, 1.5], and 2.10 just above.

(d) This follows from (c), 2.6 and a repetition of the proofs of 2.9 and 2.10 using w^*-integrals.

Before going on we give a Stonian proof of one of the most interesting results of Dunford and Pettis [1].

Corollary 2.12. If $T: L^1 \to X$ is representable and $K \subset L^1(\mu)$ is bounded and uniformly integrable, then $T(K)$ is $\| \ \|$-compact.

Proof: Since $DT \in C_o(S, X_\sigma)$, find $b_n \in A$, $b_n \nearrow e$ such that $b_n \cap N = \square$ where DT is $\| \ \|$-continuous on $S \backslash N$. Replace $g \in K \subset L^1(\mu)$ by the indefinite integral $\nu(a) = \int_A g \, d\mu$. So $\nu(e - b_n) \to 0$ uniformly over $\nu \in K$.

Let $S_n \nu = \int_{b_n} DT \, d\nu$. Since $\chi(b_n) DT \in C(S, X)$, S_n is a compact operator. So $S_n(K)$ is $\| \ \|$-compact and $\| (S_n - T)\nu \| = \| \int_{e \backslash b_n} DT \, d\nu \| \leq \int_{e \backslash b_n} \| DT \| d\nu \to 0$ uniformly over $\nu \in K$.

We conclude the section with a few remarks and some now more-or-less trivial observations. Regarding Theorem 2.8, it is not surprising on three counts that $f(S)$ should not be too large in X for $f \in C(S, X_\sigma)$. The clearest reason is that such an f defines a weakly compact operator $x^*(Tg) = \int g \cdot x^* f \, d\mu$ for $g \in L^1$ with values in X by the Krein-Smulian theorem. Hence by the classical Dunford-Pettis-Phillips Theorem, T has separable range, and then so does f, since

$f(S)$ is contained in the weak closure of $\{T(\frac{\mu_a}{\mu(a)}) : a \in A\}$. Secondly, the theory of continuous functions on products emphasizes the fact that these are often determined by countably many coordinates and S is a subspace of $\pi_A\{0, 1\}$. Thirdly we have the following trivial consequence of the Eberlein-Smulian theorem and a result of Gleason.

Theorem 2.13. If S is extremally disconnected, then no $f \in C(S, X_\sigma)$ is 1-1.

Proof: Suppose there is a 1-1 f and that $\{s_n\}$ is a sequence of distinct points in S. Then there exists $x \in f(S)$ such that $f(s_n)$ clusters weakly to $x = f(s)$ and so a subsequence $f(s_{n_k})$ converges to $f(s)$. Now f is a homeomorphism so $s_{n_k} \to s$. But by Gleason's result [4 , Thm. 1.3], no sequence in s converges unless it is ultimately constant.

These remarks along with the looseness of the proof of Theorem 2.8 (inherent in the metric d), suggests the question: If S has c.c.c. and is compact and extremally disconnected, is the conclusion of 2.8 still valid? **W. Sachermeyer [11] has communicated in the affirmative with compact and c.c.c. being sufficient.**

A second point is that although we began with only the existence in X^{**} of the limit $T(\mu_a/\mu(a))$ in the weak * topology much more is now apparent

Theorem 2.14. Let ν be a countably additive vector measure of bounded variation, and let $f(s) = w^* - \lim_{a \searrow s} \frac{\nu(a)}{\mu(a)}$. If $f \in C_o(S, X_\sigma)$ then $f(s) = || \, ||- \lim_{a \searrow s} \frac{\nu(a)}{\mu(a)}$ except on a closed nowhere dense subset of S.

Proof: Suppose, using 2.8 and 2.5, that $f(S\backslash N) \subset X$ and f is $|| \, ||$-continuous on $S\backslash N$. Given $s \in S\backslash N$ and $\epsilon>0$ choose $b \ni s$ such that $t \in b$ implies $||f(t) - f(s)|| < \epsilon$. If $s \in a \leq b$ then for $||x^*|| \leq 1$, one has

$$| x^*(f(s) - \frac{\nu(a)}{\mu(a)}) | = |x^* f(s) - \frac{x^*\nu(a)}{\mu(a)} | = | x^* f(s) - \frac{\int_a x^* f}{\mu(a)} | \text{ by 1.5 (a).}$$

Hence

$$|x^*(f(s) - \frac{\nu(a)}{\mu(a)}) | \leq \frac{1}{\mu(a)} \int_a | x^* f(s) - x^* f(t) | \, d\mu(t)$$

$$\leq \frac{1}{\mu(a)} \int_a || f(s) - f(t) || \, d\mu(t)$$

$$\leq \epsilon$$

so that $|| f(s) - \frac{\nu(a)}{\mu(a)} || \to 0$ as a s.

It should be clear how to handle the more general case, where ν is of bounded variation and countably additive but $f(s)$ is not even in X^{**} (e.g. in the case $X = \mathcal{R}$ we have $f = +\infty$ except on a nowhere dense set (1.3)). Here write $e = \bigvee_{n=1}^{\infty} a_n$, a_n with $(n\mu - |\nu|)_{a_n} \geq 0$, where $|\nu|$ is the total variation of ν. Then one uses $f \in C_o(S, X_\sigma)$ where for each n, $f_n(s) = \chi(a_n)f$.

A final question yet remains. Note that even though $f \in C(S, X_\sigma)$ is $\|\ \|$-continuous on some $S\backslash N$, one still has $f(S) \subset X$. In 2.13 we see that the limit of $\nu(a)/\mu(a)$ exists in the norm at points of $\|\ \|$-continuity, a weaker conclusion. Is it true that if the $\lim_{a \searrow s} \frac{\nu(a)}{\mu(a)} \in X$ for all s, then the limit exists at all s in the norm, at least for X with RNP? This would appear to be the proper analogue of 1.2 for ν X-valued. That is, is a weakly compact operator $\|\ \|$-differentiable at every $s \in S$?

Finally using 2.2(c), the remarks in the proof of 2.8, and using the fact that $\| T \| = \| DT \|$, we have from 2.11 that $C(S, X_\sigma)$ $(C(S, X))$ is isometrically isomorphic to the Banach space $\mathcal{L}_w(L^1, X)$ $(\mathcal{L}_c(L^1, X))$ of weakly compact (compact) maps of L^1 into X. Pryce [9 , 3.2, 3.4] shows that, $C(S, X_\sigma)$ is angelic in the topology of pointwise convergence at least when $\sigma(X, X^*)$ is metrizable, and, compact and relatively countably compact sets in $C(S, X_\sigma)$ coincide in any case. This raises the obvious question as to when these properties hold for $\mathcal{L}_w$ and $\mathcal{L}_c$.

3. <u>Lifting and the point values of $D_\mu\nu$</u>.

In this section Ω denotes the interval (-1, 1) and m denotes Lebesgue measure on the Lebesgue measurable sets $\mathcal{L}$ as well as on the Boolean algebra $A = \mathcal{L}/m^{-1}(0)$. We now try to relate the point values $D_m\nu(s)$ at $s \in S$ to the function values $f(x)$ where $\nu(a) = \int_A f(x)\, dm(x)$, $A \in \mathcal{L}$, for most $s \in S$ and $x \in \Omega$.

The difficulty is of course the problem of lifting $L^\infty(\Omega, m)$, originally solved by Von Neumann years ago. Our study is quite simple and rests on two points: (1) Ω maps 1-1 into the Stone space $S_\mathcal{L}$ of $\mathcal{L}$ by the map $\theta(x)(E) = \chi_E(x)$, but the Stone space S of $\mathcal{L}/m^{-1}(0)$ is a closed subset of $S_\mathcal{L}$ disjoint from $\theta(\Omega)$ (2) We remedy this by mapping points of Ω into certain closed nowhere dense subsets of S and on each of these we show $D_m \nu$ is constant a.e. on Ω.

While the ensuing study applies to any $\nu \in L^1(\Omega, m)$ qua a measure on $L/m^{-1}(0)$, it is plain from 1.3 that it suffices to deal with measures $\nu(E) = \int_E f\,d\mu$, for $[f] \in L^\infty(\Omega, L, m)$. Since $\hat{f}$ is continuous (Sec. 1) and $\nu(a) = \int_a \hat{f}\,dm$ it follows without recourse to even 1.2 that $D_m\nu(s) = \hat{f}(s)$, so in fact all we will study are the point values $\hat{f}(s)$.

To begin we denote by $d(x, A)$ the limit

$$\lim_{\delta \to 0} \frac{m(A \cap \Delta)}{m(\Delta)}$$

where $\Delta = (x - \delta, x + \delta)$, whenever it exists. If $a \in A$ is $A + m^{-1}(0)$ we set $d(x, a) = d(x, A)$; $d(x, A)$ is called the density of A at x.

Initally we assume but one result (which we have tried to obtain from 1.2 without success) a weakened form of the Lebesgue Density Theorem [8 , 3.20]:

Theorem 3.1. (Lebesgue) If $a \neq 0$, there exists $x \in \Omega$ such that $d(x, a) = 1$.

It follows that if $a \neq e$ there exists x such that $d(x, a) = 0$.

Fix $x \in X$ and define the ideals

$I_x = \{a \in A\colon d(x, a) = 0\}$

$J_x = \{a \in A\colon m(a\Delta) = 0 \text{ for some } \Delta \ni x\}$

Since $A \in m^{-1}(0)$ implies $d(x, A) = 0$ and $m(A \cap \Delta) = 0$, both I_x and J_x are legitimately ideals in the quotient algebra A, and, $I_x \subsetneq J_x \subsetneq A$.

We make the notational convention of always writing Δ for an interval $(x - \delta, x + \delta)$ centered about a point x under discussion and also write Δ for the clopen version of $\Delta + m^{-1}(0)$ in S. Also throughout this section we choose to interpret points $s \in S$ as 0 - 1 valued Boolean homomorphisms of A onto $\{0, 1\}$ where, $1 + 1 = 0$.

Dual to the ideals I_x and J_x are the closed sets

$$C_x = \{s \in S\colon \; Ix \subset s^{-1}(0)\}$$

and $$D_x = \cap\, \{\Delta\colon x \in \Delta\}.$$

Let T^0 denote the interior of a set $T \subset S$.

Lemma 3.2. $C_x \subsetneq D_x$ and D_x is nowhere dense.

Proof: Since in Ω, $(x - \delta, x + \delta) \searrow \{x\}$, then in A, $\Delta \searrow 0$. Hence $D_x^0 = \square$, and D_x

is closed, being an intersection of closed sets. Now C_x is clearly closed (since the Stone topology is pointwise convergence on A: $s_\alpha(a) \to s(a)$ for all $a \in A$) and since $J_x \subset I_x$, then $C_x \subset D_x$. Hence $C_x^0 = \square$.

Now choose $b \in I_x \backslash J_x$ and let K be the ideal $J_x + b\,A \subset I_x$. If $s^{-1}(0) \supset K$, then $s(b) = 0$ so $s \notin b$. But, then $s^{-1}(0) \supset J_x$ and $b \supset C_x$ and hence $s \in D \backslash b$; or, $s \in D_x \backslash C_x$.

The idea of this section is to map Ω into S via $x \to C_x$ or $x \to D_x$. As things turn out D_x is too large, but C_x suffices. If we recall for a moment the alternate interpretation of points $s \in S$ as maximal ideals $s^{-1}(0) \subset A$, we see just what C_x is: the collection of maximal ideal extensions of the ideal I_x of sets with density 0 at x.

We pause to clear up a technical difficulty, but also one which limits our study to the real line. This is probably well-known.

<u>Lemma 3.3</u>. Let F be any filter of open intervals about x such that $\cap F = \{x\}$. Let ν be any non-negative measure on L, $\nu \ll m$ and let $f(x) = \nu(-1, x)$. These are equivalent:

$$(1) \quad \lim_{\Delta \searrow x} \frac{\nu(\Delta)}{m(\Delta)} = 0$$

$$(2) \quad f'(x) = 0$$

$$(3) \quad \lim_{F \searrow x} \frac{\nu(F)}{m(F)} = 0 \qquad (F \in F)$$

In (1), Δ has the form $(x - \delta, x + \delta)$.

<u>Proof</u>: For $\Delta = (x - \delta, x + \delta)$, let $\Delta^+ = (x, x + \delta)$.

Suppose (1) and $\overline{\lim_{\Delta^+ \searrow x}} \frac{\nu(\Delta^+)}{m(\Delta^+)} \geq \alpha > 0$. So there exists Δ such that

(a) $\frac{\nu(\Delta)}{m(\Delta)} < \alpha/4$ and (b) $\frac{\nu(\Delta^+)}{m(\Delta^+)} > \alpha/2$. Hence from (b),

$$\nu(\Delta^+) - \alpha/2\ m(\Delta^+) > 0$$

or

$$\nu(\Delta^+) - \alpha/4\ m(\Delta^+) > \frac{\alpha}{4} m(\Delta^+)$$

and by (a)

$$\begin{aligned} 0 &> \nu(\Delta) - \frac{\alpha}{4} m(\Delta) \\ &> \nu(\Delta^+) - \frac{\alpha}{4} m(\Delta^+) + \nu(\Delta^-) - \frac{\alpha}{4} m(\Delta^-) \\ &> \frac{\alpha}{4} m(\Delta^+) + \nu(\Delta^-) - \frac{\alpha}{4} m(\Delta^-) \\ &> \nu(\Delta^-), \text{ a contradiction} \end{aligned}$$

Thus, $f'_+(x) = 0$ and similarly, $f'_-(x) = 0$. Hence (1) implies (2).

Given (2), there exists $\delta > 0$ such that $\delta \leq \delta_o$ implies

$$\frac{\nu(\Delta^+)}{m(\Delta^+)} < \epsilon \text{ and } \frac{\nu(\Delta^-)}{m(\Delta^-)} < \epsilon$$

Given F, pick $I_o = (x - \delta', x + \delta'') \in F$ so that $\delta', \delta'' < \delta_o$.

Then $\nu(x, x + \delta'') < \epsilon\, m(x, x + \delta'')$

$\nu(x - \delta', x) < \epsilon\, m(x - \delta', x)$

whence $\nu(I_o) < \epsilon\, m(I_o)$, or $\lim_{F \downarrow x} \frac{\nu(I)}{m(I)} = 0$.

Hence (2) implies (3) and clearly (3) implies (1).

Note that 3.3 applied to $\nu(E) = m(A \cap E)$ says that I_x, and hence C_x, is independent of any particular neighborhood filter of intervals about x, not necessarily centered at x.

<u>Theorem 3.4</u>. $d(x, a) = 1$ iff $C_x \subset a$.

<u>Proof</u>: If $d(x, a) = 1$ then $d(x, e - a) = 0$ so, for all $s \in C_x$, $s(e - a) = 0$ or $s(a) = 1$ so $s \in a$. Hence $C_x \subset a$.

Conversely, suppose $C_x \subset a$ and $d(x, a) \neq 1$. Let $M_x = I_x + a A$. Then M_x is an ideal containing I_x and $M_x \neq A$. For if $e = c + ab$, $c \in I_x$, then $ab = e - c$ and hence $d(x, ab) = 1$ since $d(x, c) = 0$. But $1 \geq d(x, a) \geq d(x, ab) = 1$ so $d(x, a) = 1$, a contradiction. So $M_x \neq A$ and there exists $s \in S$ such that $M_x \subset s^{-1}(0)$. But then, $I_x \subset s^{-1}(0)$ so $s \in C_x$. But now $s(a) = 0$ whence $s \notin a$ yet $C_x \subset a$, a contradiction.

<u>Corollary 3.5</u>. $a = \overline{\cup \{C_x\colon C_x \subset a\}}$

<u>Proof</u>: If not there is a $b \subset a$ such that $b \cap \cup \{C_x\colon C_x \subset a\} = \square$. But by 3.1 there exists $x \in \Omega$ such that $d(x, b) = 1$, whence also $d(x, a) = 1$, a contradiction.

<u>Corollary 3.6</u>. (1) $d(x, a) = 0$ iff $a \cap C_x = \square$

(2) $C_x = \cap \{a\colon d(x, a) = 1\}$

(3) $D_x = \cap \{a\colon \exists\ \Delta \ni x \text{ such that } m(a\, \Delta) = m(\Delta)\}$

Proof: (1) and (2) follow from 3.5. For (3), if $m(a\Delta) = m(\Delta)$ then $m((e - a)\Delta) = 0$ so $e - a \in J_x$. Hence $s \in D_x$ implies $s(a) = 1$. Hence $D_x \subset a$. Conversely,

$$\cap \{a: \exists \Delta \ni x, \ni m(a\Delta) = m(\Delta)\}$$

$$\subset \cap \{\Delta: x \in \Delta\} = D_x \text{ by definition.}$$

Remark 3.7. (a) No C_x is a singleton. Fix $\Delta \ni x$, write $\Delta = (x - \delta, x) \cup [x, x + \delta), \Delta^- \cup \Delta^+$. Now there is an s such that $s(\Delta) = 1$ hence one of $s(\Delta^-)$, or $s(\Delta^+)$ is 0. If $s(\Delta^-) = 0$ then since $s(e - \Delta) = 0$, then $s(\Delta_1^-) = 0$ for all $\Delta_1 \leq \Delta$, and, $s(\Delta_1^+) = 1$. Now let $t \in S$, $t^{-1}(0) \supset I_x + \Delta^- \cdot A$. Then $t \in C_x$ and $t \neq s$.

(b) Let F be a filter of open intervals about x such that $\cap F = \{x\}$. Then $\{b: B \in F\}$ is cofinal at no $s \in D_x$ (in fact, no $s \in S$). For, choose $b \in A$ so that $d(x, b) \neq 0$ or 1. Then $\lim_{a \searrow s} \frac{m_b(a)}{m(a)} = \chi(b)$ is 0 or 1, so F cannot be cofinal at s, using 3.3.

Lemma 3.8. $x \neq y$ in Ω iff $C_x \cap C_y = \square$.

Proof: $x \neq y$ implies there exists Δ_x, Δ_y such that $x \in \Delta_x$, $y \in \Delta_y$ and $\Delta_x \cap \Delta_y = \square$. Hence $D_x \cap D_y = \square$ so $C_x \cap C_y = \square$. Conversely, if $C_x \cap C_y = \square$ there exists a, b such that $a \supset C_x$, $b \supset C_y$, and $ab = \square$. So $d(x, a) = 1$, yet $d(y, a) = 0$ so $x \neq y$.

The class $C = \{C_x: x \in \Omega\}$ is almost a decomposition of S, for $S = e = \overline{\bigcup_{x \in \Omega} C_x}$. Moreover it is clear that for some a, $a \supseteq \cup\{C_x: C_x \subset a\}$; for example, if $a = \text{clopen } [\frac{1}{2}, 1)$, then $C_{1/2} \cap a \neq \square$ and $C_{1/2} \cap (e - a) \neq \square$ and $C_{1/2} \cap C_x = \square$ for any $x \neq 1/2$, where $C_x \subset a$.

Now 3.8 also gives a simple argument that C_x is nowhere dense. For if there is an $a \subset C_x$, then there is a y such that $d(y, a) = 1$ whence $C_y \subset a \subset C_x$ so $C_y = C_x$ on $a = C_x$. But then a has exactly one point of density 1, an impossibility.

The preceeding discussion is mainly preliminary; the next result is the key to this section and to a simple proof of the lifting theorem. We shall see that if $[f] \in L^\infty(m)$, then $\hat{f}$ is constant on C_x for almost all x. To accomplish this we now need the full strength of the Lebesgue density theorem: If $A \in L$ and $A_1 =$

$\{x: d(x, A) = 1\}$ then $A_1 \in L$ and $m(A \setminus A_1) = m(A_1 \setminus A) = 0$. That is, $a = a_1$.

Question: That $A_1 \in L$ is easy. Does $m(A \setminus A_1) = m(A_1 \setminus A) = 0$ follow from 3.5?

It suffices to consider an equivalence class $[f] \in L^{\infty}(X)$ such that $1 \geq f_o \geq 0$ a.e. for all $f_o \in [f]$. So we can suppose $0 \leq f \leq 1$ everywhere. (Corresponding to $[f]$ is of course a single function $\hat{f} \in C(S)$). The key idea is

Lemma 3.9. Let $\epsilon > 0$ and let $A_\epsilon = \{x \in X$: there exists $s \in C_x$ such that $\hat{f}(s) > f(x) + \epsilon\}$. Then $m(A_\epsilon) = 0$.

Proof: Suppose $m^*(A_\epsilon) > 0$. Find $B \in L$ such that $B \supset A_\epsilon$ and $m^*(A_\epsilon) = m(B)$ and let $B_1 = \{x \in X: d(x, B) = 1\}$. Then $B_1 \in L$ and $m(B_1 \Delta B) = 0$. Note that if $x \in B_1$, then $b_1 \supset C_x$ by 3.6.

Consider $\chi_{B_1} f$. We know that $0 < || \chi_{B_1} f ||_\infty = || \chi(b_1) \hat{f} ||$. Let $M = (|| \chi(b_1) \hat{f} || - \epsilon) \vee 0$ and let $E = \{x: (\chi_{B_1} f)(x) > M\}$, so that $m(E) > 0$ and $E \subset B_1$.

Suppose there exists $x \in E \cap A_\epsilon$. Then $x \in B_1$ as well so that (1) $C_x \subset b_1$ and (2) there exists $s \in C_x$ such that $\hat{f}(s) > f(x) + \epsilon$. Hence $f(x) < \hat{f}(s) - \epsilon \leq \sup_{t \epsilon b_1} \hat{f}(t) - \epsilon \leq M$. But $x \in E$ yields $f(x) > M$. So $E \cap A_\epsilon = \square$. Hence,

$$\begin{aligned} m^*(A_\epsilon) &= m(B) = m(B_1) \\ &= m(E) + m(B_1 \setminus E) \\ &= m(E) + m(B \setminus E) \text{ since } m(B \Delta B_1) = 0 \\ &= m(E) + m^*(B \setminus E) \\ &\geq m(E) + m^*(A_\epsilon) \end{aligned}$$

since $A_\epsilon \subset B$ and $A_\epsilon \cap E = \square$. This is a contradiction, since $m(E) > 0$.

Corollary 3.10. $\hat{f}$ is constant on C_x, for almost all x.

Proof: Let $A_n = \{x: \exists\, s \in C_x \ni \hat{f}(s) > f(x) + \frac{1}{n}\}$

Then $m(A_n) = 0$ and hence,

$$A^+ = \{x: \exists\, s \in C_x \ni \hat{f}(s) > f(x)\}$$

has measure 0. Similarly, $A^- = \{x: \exists\, s \in C_x \ni \hat{f}(s) < f(x)\}$ has zero measure. Hence

$$m\{x: \exists\, s \in C_x \ni \hat{f}(s) \neq f(x)\} = 0.$$

So for all x not in $A^+ \cup A^-$, $\hat{f}(s) = f(x)$ for all $s \in C_x$. Q.E.D.

By a selection we will mean a map $\rho: \Omega \to \bigcup_{x \in \Omega} C_x$ such that $\rho(x) \in C_x$.

Given ρ and $\hat{f} \in C(S)$ let $f_\rho(x) = \hat{f}(\rho(x))$.

<u>Corollary 3.11</u>. $f_\rho \in [f]$.

<u>Proof</u>: For almost all x, $f_\rho(x) = \hat{f}(\rho(x)) = f(x)$ by 3.10.

<u>Corollary 3.12</u>. The map $L_\rho : L^\infty(m) \to R^\Omega$ given by $L_\rho[f] = f_\rho$ is a lifting of $L^\infty(m)$.

<u>Proof</u>: This is obvious because the map $[f] \to \hat{f}$ preserves sums, products and scalar multiples and $L_\rho[f] \in [f]$ for every class [f].

A lifting L is called a <u>strong lifting</u> if $L[f](x) = f(x)$ for all x, for all continuous f: Let $C_b(X)$ denote the bounded continuous functions on X.

<u>Theorem 3.13</u>. Any lifting L_ρ is a strong lifting. Indeed, if $f \in C_b(\Omega)$, then $\hat{f}|C_x = f(x)$ for all x.

<u>Proof</u>: In fact we will (unfortunately!) show that $\hat{f}|D_x = f(x)$. Suppose there is an $s \in D_x$ such that $\hat{f}(s) > f(x) + \epsilon$. Then there exists $\delta > 0$ such that $t \in (x - \delta, x + \delta)$ implies $f(t) < \hat{f}(s) - \epsilon$. Let a_o = clopen $(x - \delta, x + \delta)$ in S. Then $D_x \subset a_o$ and if $s \in a \subset a_o$, then defining $\nu(b) = m(\chi(b)\hat{f})$ for all b, we have

$$\frac{\nu(a)}{m(a)} = \frac{1}{m(a)} \int_a \hat{f}\, dm = \frac{1}{m(A)} \int_A f\, dm.$$

Hence
$$\frac{\nu(a)}{m(a)} \leq \frac{1}{m(a)} \int_a (\hat{f}(s) - \epsilon)\, dm = \hat{f}(s) - \epsilon$$

since $A \sim a \leq a_o$ implies $f \leq \hat{f}(s) - \epsilon$ a.e. on A - that is, since $m(A \cap ((x - \delta, x + \delta) \Delta A)) = 0$. But $\lim_{a \searrow s} \frac{\nu(a)}{m(a)} = \hat{f}(s)$, a contradiction. Hence $\hat{f}|D_x \leq f(x)$ and similarly $f|D_x \geq \hat{f}(x)$. Thus L_ρ is a strong lifting because $L_\rho[f](x) = \hat{f}(\rho(x)) = f(x)$ for all x.

It is unfortunate that $\hat{f}|D_x$ is constant and not merely constant on C_x, because, since it is the C_x's that prove the lifting theorem, one would like to define these for more general spaces X and obviously now $C_b(X)$ will not suffice. Since $C_b(X)$ is a proper subalgebra of $L^\infty(X) = C(S)$, the weak topology of $C_b(X)$ on S is not T_2. In fact, $C_b(\Omega)$ cannot distinguish any points of any set D_x. R. F. Wheeler [15] has recently related these ideas to the projective resolution $E(\Omega)$ of Ω with the density topology and shown that $D_x \cap E(\Omega)$ is C_x, and that 3.13 holds for the functions f continuous in the density topology on Ω.

There are now obviously many liftings L_ρ of L^∞. In one sense however they are all the same.

Theorem 3.14. Let ρ be any selection and let X_ρ be X with the ρ-weak topology. Then X_ρ is homeomorphic to a dense subset of S and $\beta X_\rho \cong S$.

Proof: Because $a \in \mathcal{A}$ implies there is an x with d(x, a) = 1, whence $a \supset C_x$, every a hits $\rho(X)$. So X_ρ is a homeomorphic to a dense subset of S.

Hence ρ uniquely extends to a continuous onto map $\rho\colon \beta X_\rho \to S$. Since S is compact and extremely disconnected we can, by using a result of Gleason [4], show that ρ is a homeomorphism if we can show $\rho(E) \neq S$ for any proper closed subset $E \subset \beta X_\rho$.

If E is closed and proper, we cannot have $E \supset X$. Choose $x \in X\setminus E$. Since E is closed there is a clopen $a \subset S$ such that $x \in \rho^{-1}(a)$ and $\rho^{-1}(a) \cap E = \Box$. Whence $a \cap \rho(E) = \Box$ so $\rho(E)$ is proper (and closed).

Corollary 3.15. If $\bar{\rho}\colon \beta X_\rho \to S$ is the unique extension of ρ to βX_ρ, then $\bar{\rho}^{-1}(C_x) \cap X = \{x\}$. Thus $\beta X_\rho \setminus X_\rho = \bar{\rho}^{-1}(S\setminus\rho(X))$.

Proof: If not, then there exists $y \in X \cap \bar{\rho}^{-1}(C_x)$, $y \neq x$ whence

$$\rho(y) \in C_x \cap C_y \text{ so } x = y \text{ by 3.8.}$$

A selection ρ is quite arbitrary. In fact, there are special selections obtained as follows. Pick $s_o \in C_o = \cap\{a\colon d(0, a) = 1\}$. Let $s_x(a) = s_o(a - x)$ where a - x is the clopen set obtained from A - x. Then $s_x \in C_x$ for if d(x, a) = 0, then d(0, a - x) = 0 whence $s_x(a) = s_o(a - x) = 0$ since $a - x \in I_o$. Then $\rho_o(x) \equiv s_x$ yields the kind of lifting obtained by Scheinberg [12].

We now want to relate the previous results to point values at $x \in \Omega$ and to differentiation of the integral on the line. Here we let $m_f(A) \equiv \int_A \hat{f}\, dm \equiv m_f(a) = \int_a \hat{f}\, dm$, the last integral being in the Stone space by the induced Lebesgue measure. It is well known that $\frac{dm_f}{dm} \equiv \lim_{\Delta \searrow x} \frac{m_f(\Delta)}{m(\Delta)} = f(x)$ a.e. indeed, that $\frac{dm_f}{dx} \equiv \lim_{\delta \searrow 0} \frac{m_f(x, x \pm \delta)}{\delta} = f(x)$ a.e. These need not coincide, viz. $f = \chi_{(1/2,\, 1)}$ at $x = 1/2$. What is perhaps slightly less well-known is that

$$\lim_{\Delta \searrow x} \frac{1}{m(\Delta)} \int_\Delta |f - f(x)|\, dm = 0$$

a.e. x, and here Δ need not be centered at x by 3.3. The set of x for which this holds is known as the Lebesgue set of f. We will show that this is essentially the set of those x for which $\hat{f}|C_x$ is constant.

Let a and b be given, k is constant. Then

$$\int_b |\hat{f} - k| \le \int_b |\hat{f} - \hat{f}\,\chi(ab)| + \int_b |\hat{f}\,\chi(ab) - k\,\chi(ab)| + \int_b |k\,\chi(ab) - k|$$

$$\le \int_{b\backslash ab} |\hat{f}| + \int_{ab} |\hat{f} - k| + \int_{b\backslash ab} |k|$$

$$\le m(b\backslash ab)(\|\hat{f}\| + k) + \int_{ab} |\hat{f} - k|$$

Theorem 3.15. If $\hat{f}|C_x = k$, then $\epsilon>0$ implies there exists an interval $\Delta_0 \ni x$ such that if $x \in \Delta \subset \Delta_o$, then $\frac{1}{m(\Delta)} \int_\Delta |f - k|dx < \epsilon$. In particular, $\frac{dm_f}{dx}(x) = k$

Proof: In the above inequality, replace b by Δ and divide both sides by $m(\Delta)$ to obtain

$$\frac{1}{m(\Delta)} \int_\Delta |\hat{f} - k| \le \frac{m(\Delta\backslash a\Delta)}{m(\Delta)} (\|\hat{f}\| + k) + \frac{1}{m(\Delta)} \int_{a\Delta} |\hat{f} - k|$$

$$\le \left[1 - \frac{m(a\Delta)}{m(\Delta)}\right] (\|\hat{f}\| + k) + \frac{1}{m(\Delta)} \int_{a\Delta} |\hat{f} - k|$$

Given $\epsilon>0$ first choose $a \supset C_x$ so that $|\hat{f} - k| < \epsilon$ on a. Then, since $a \supset C_x$, whence $d(x, a) = 1$, choose Δ_o such that $x \in \Delta \le \Delta_o$ implies $\left|1 - \frac{m(a\Delta)}{m(\Delta)}\right| < \epsilon/(\|\hat{f}\| + k)$ to obtain $\frac{1}{m(\Delta)} \int_\Delta |f - k| = \frac{1}{m(\Delta)} \int_\Delta |\hat{f} - k| < 2\epsilon$ since $a\Delta \subset a$. Since Δ is any open interval about x the rest is clear by 3.3.

Note that by 3.3, Δ need not be centered at x in 3.15.

Theorem 3.16. If $\lim_{\Delta \searrow x} \frac{1}{m(\Delta)} \int_\Delta |f - k|\, dm = 0$, then $\hat{f}|C_x = k$.

Proof: Suppose $s \in C_x$ and $\hat{f}(s) \ne k$. Given $\epsilon < |\hat{f}(s) - k|$ find $a \ni s$ such that $|\hat{f} - k| > \epsilon$ on a. Now $a \cap C_x \ne \square$ so $d(x, a) \ne 0$ and there exists $\alpha > 0$ and a sequence of intervals Δ_n x such that $\frac{m(a\Delta_n)}{m(\Delta_n)} \ge \alpha$ for all n. Hence

$$\frac{1}{m(\Delta_n)} \int_{\Delta_n} |f - k|dm = \frac{1}{m(\Delta_n)} \int_{\Delta_n} |\hat{f} - k|dm$$

$$\ge \frac{1}{m(\Delta_n)} \int_{a\Delta_n} |\hat{f} - k|dm$$

$$\ge \frac{m(a\Delta_n)}{m(\Delta_n)} \epsilon$$

$$\ge \epsilon\, \alpha > 0$$

Remark: 3.16 gives an alternate proof of 3.10, given the theorem that the Lebesgue set of f has full measure. In fact more is true. There is a Lebesgue set for the class [f] namely, $\{x: \hat{f}|C_x$ is constant$\}$, which contains the Lebesgue set for any $g \in [f]$.

Note also that absolute value is essential in 3.16: Consider $f = \chi_A$ at x for which $d(x, A) \neq 0$ or 1.

Theorem 3.17. If $\hat{f}|D_x = k$, then for any $g \in [f]$, g is nearly continuous at x: $\epsilon>0$ implies there exists $\delta > 0$ such that for almost all $t \in (x - \delta, x + \delta)$ one has $|g(t) - k| < \epsilon$ for all $g \in [f]$, with δ independent of g.

Proof: Given $\epsilon>0$ find $a_o \supset D_x$ so that $|\hat{f}(s) - k| < \epsilon$ for $s \in a_o$. Now $D_x = \cap \{\Delta: x \in \Delta\}$ and there exists Δ_o such that $\Delta_o \subset a_o$. For if $\Delta \backslash a_o \neq \Box$ for all Δ, then by compactness of $e - a_o$, $\underset{\Delta}{\cap}(\Delta\backslash a_o) \neq \Box$, yet $\underset{\Delta}{\cap}(\Delta\backslash a_o) = D_x\backslash a_o = \Box$. Write $(x - \delta, x + \delta) = \Delta_o \subset a_o$. Then given $g \in [f]$, for almost all $t \in \Delta_o$, $\hat{f}|C_t$ is constant and equal to $g(t)$, so $|g(t) - k| = |\hat{f}|C_t - k| < \epsilon$, since $t \in \Delta_o$ implies $C_t \subset D_t \subset \Delta_o \subset a_o$.

Theorem 3.18. Suppose k is a number such that 0 is a cluster point of the net $\{\frac{1}{m(\Delta)} \int_\Delta |f - k|\}_{\Delta \searrow x}$. Then there is an $s \in C_x$ such that $\hat{f}(s) = k$.

Proof: If not, by compactness of C_x, there is an $a \supset C_x$ such that $|\hat{f} - k| \geq \epsilon$ on a. Then

$$\frac{1}{m(\Delta)} \int_\Delta |f - k| \geq \frac{1}{m(\Delta)} \int_{a\Delta} |\hat{f} - k| \geq \epsilon \frac{m(a\Delta)}{m(\Delta)}$$

But since $a \supset C_x$, $\frac{m(a\Delta)}{m(\Delta)} \to 1$, yielding a contradiction.

I have no proof for a converse. What is needed is that, if $d(x, a) \neq 0$, then there is a sequence Δ_n of decreasing intervals such that $x \in \bar{\Delta}_n$ and such that $m(a\cdot\Delta_n)/m(\Delta_n) \to 1$. Then the same proof as 3.15 applies.

Our final result concerns Bochner integrable functions f, though as before we consider only $f \in L^\infty(\Omega,X,L,m)$ for simplicity. It is known [1 , p. 49] that

(*) $\lim\limits_{\Delta \searrow \omega} \frac{1}{m(\Delta)} \int_\Delta ||f(t) - f(\omega)|| \, dm(t) = 0$ for almost all $\omega \in \Omega$.

Theorem 3.19. If $f \in L^\infty(\Omega,X,L,m)$, and (*) above holds at ω, then $\hat{f}|C_\omega = f(\omega) \in X$. For almost all ω, $\hat{f}$ is $|| \; ||$-continuous on C_ω and if

$\hat{f}|C_\omega = x$ and $\hat{f}$ is $\|\ \|$-continuous on C_ω, then (*) holds with $f(\omega)$ replaced by x.

Proof: Since $\hat{f}$ is $\|\ \|$-continuous on $S\backslash N$, where N is closed and nowhere dense, if we write $S\backslash N = \bigcup_{n=1}^{\infty} b_n$, $b_n \in A$, then $\hat{f}$ is $\|\ \|$-continuous on any C_{ω_o} with $\omega_o \in \bigcup_{n=1}^{\infty} \{\omega: d(\omega, b_n) = 1\}$, which has full measure. A repeat of the proof of 3.15, where $\hat{f}|C_\omega = x$ replaces k, completes the proof of the second statement in the theorem.

To prove the first statement, first suppose $h \in L^\infty(\Omega, X, L, m)$ and let $g(\omega) = \|h(\omega)\|$. Then $g \in L^\infty(\Omega, m)$ since h is strongly measurable [1, II, 1.1]. For $A \in L$, and all ω, $g(\omega)\chi_A(\omega) = \|h(\omega)\chi_A(\omega)\|$. Hence $\|g\chi_A\|_\infty \geq \|h(\omega)\chi_A(\omega)\|$ for all ω and therefore, $\|g\chi_A\|_\infty \geq \|h\chi_A\|_\infty$.

Now suppose $\hat{g}(s) < \|\hat{h}(s)\|$ for some $s \in S$. Since $\hat{g} \in C(S)$ and $s \to \|\hat{h}(s)\|$ is l.s.c. (see the proof of 2.4), there is an $a \ni s$ such that $\hat{g}(t)\chi(a)(t) < \|\hat{h}(t)\chi(a)(t)\|$ for all $t \in a$. Thus $\hat{g}(t)\chi(a)(t) < \|\chi(a)\hat{h}\|$ for all t. But this means $\|\chi_A g\|_\infty = \|\chi(a)\hat{g}\| < \|\chi(a)\hat{h}\| = \|\chi_A h\|_\infty$ by 2.2 (c). This is a contradiction. Hence, $\hat{g}(s) \geq \|\hat{h}(s)\|$ for all $s \in S$.

Now suppose $\lim_{\Delta \searrow \omega} \frac{1}{m(\Delta)} \int_\Delta \|f(\alpha) - f(\omega)\| dm(\alpha) = 0$. We will show that $\hat{f}|C_\omega = f(\omega)$. If not, then there is an $s_o \in C_\omega$ such that $0 < \epsilon < \|\hat{f}(s_o) - f(\omega)\|$ and because this last function is l.s.c. in s, there is an $a \ni s_o$ such that $\epsilon < \|\hat{f}(t) - f(\omega)\|$ for all $t \in a$.

Let $h = f - f(\omega)$ and $g(\alpha) = \|h(\alpha)\|$ for $\alpha \in \Omega$.

Then $\int_\Delta \|f(\alpha) - f(\omega)\| dm(\alpha) = \int_\Delta g(\alpha)\, dm(\alpha) = \int_\Delta \hat{g}(s)\, dm(s)$

$\geq \int_\Delta \|\hat{h}(s)\| dm(s) = \int_\Delta \|\hat{f}(s) - f(\omega)\| dm(s)$

So $\frac{1}{m(\Delta)} \int_\Delta \|f - f(\omega)\| dm \geq \frac{1}{m(\Delta)} \int_{a\Delta} \|\hat{f}(s) - f(\omega)\| dm(s)$.

Since $s_o \in C_\omega$ and $s_o \in a$, then $d(\omega, a) \neq 0$ and if we then choose $\Delta_n \searrow \omega$ as in the proof of 3.16, we obtain

$$\frac{1}{m(\Delta_n)} \int_{\Delta_n} \|f - f(\omega)\| dm \geq \frac{m(a\Delta_n)}{m(\Delta_n)} \epsilon$$

with the last term bounded away from zero, a contradiction.

Corollary 3.20. Any selection $\rho: \Omega \to \bigcup_{x \in \Omega} C_x$ defines a lifting $[f] \to \hat{f} \circ \rho$ of $L^\infty(\Omega,X,L,m)$ where $\hat{f} \circ \rho \in X$ a.e.m.

In closing it seems to the author that the fact that a measure on an algebra can be differentiated at every point of S has not really been exploited. A point $s \in S$ being a 0-1 homomorphism of A is a true-false statement about elements $a \in A$ such that $\{a \in A: s(a) \text{ is "false"}\}$ (that is, $s^{-1}(0)$) is a maximal ideal. But any theorem about elements of an algebra is a true-false statement until proven always true, or false for one element. What kind of a "theorem" about a measure on an algebra A, or about elements of the algebra, can, if assumed false for at least one element, be then interpreted as a single existing point s in S? If some such hoped for "theorem" can be so transformed, then the fact that measures can be differentiated at the very point s, can, if this leads to some contradiction, prove the theorem.

Bibliography

1. J. Diestel and J. J. Uhl, Vector Measures, A.M.S. Surveys, No. 15, Providence, 1977.

2. N. Dunford and J. Schwartz, Linear Operators, Part I, Interscience New York, 1958.

3. D. H. Fremlin, Topological Riesz Spaces and Measure Theory, Cambridge University Press, London, 1974.

4. A. M. Gleason, Projective topological spaces, Illinois J. Math. 2 (1958) 482-89.

5. W. H. Graves, On the theory of vector measures, Amer. Math. Soc. Mem. #195 (Nov. 1977).

6. W. H. Graves and D. Sentilles, Completion of the universal measure: the dual of the space of measures, J. Math. Annal. and Applices, 68 #1 (1979) 228-64.

7. P. R. Halmos, Lectures on Boolean algebras, Van Nostrand, Princeton, 1963.

8. J. C. Oxtoby, Measure and category, Springer, New York, 1971.

9. J. D. Pryce, A device of R. J. Whitley applied to pointwise compactness in spaces of continuous functions, Proc. London Math. Soc., (3) (1971) 532-46.

10. A. P. Robertson and W. P. Robertson, Topological vector spaces, Cambridge University Press, New York, 1964.

11. W. Sachermeyer, On a paper of D. Sentilles, preprint.

12. S. Scheinberg, Topologies which generate a complete measure algebra, Adv. in Math. 7 (1971), 231-239.

13. D. Sentilles, An L^1-space for Boolean algebras and semi-reflexivity of spaces $L^\infty(X,\Sigma,\mu)$, Trans. Amer. Math. Soc., 226 (1977), 1-37.

14. R. Sikorski, Boolean algebras, Springer-Verlag, Berlin, 1960.

15. R. F. Wheeler, Topological Measure theory for completely regular spaces and their projective covers, Pacific J. Math. 82, #2 (1979) 565-84.

University of Missouri
Columbia, MO. 65211
January, 1979